Texts and
Monographs
in Physics

R.G. Newton

Inverse Schrödinger Scattering in Three Dimensions

Springer-Verlag
Berlin Heidelberg New York
London Paris Tokyo Hong Kong

Professor Roger G. Newton

Department of Physics, Indiana University,
Swain Hall-West 117, Bloomington, Indiana 47405, USA

Editors

Wolf Beiglböck
Institut für Angewandte Mathematik
Universität Heidelberg
Im Neuenheimer Feld 294
D-6900 Heidelberg 1
Fed. Rep. of Germany

Elliott H. Lieb
Department of Physics
Joseph Henry Laboratories
Princeton University
Princeton, NJ 08540, USA

Joseph L. Birman
Department of Physics, The City College
of the City University of New York
New York, NY 10031, USA

Tullio Regge
Istituto di Fisica Teorica
Università di Torino, C. so M. d'Azeglio, 46
I-10125 Torino, Italy

Robert P. Geroch
Enrico Fermi Institute
University of Chicago
5640 Ellis Ave.
Chicago, IL 60637, USA

Walter Thirring
Institut für Theoretische Physik
der Universität Wien, Boltzmanngasse 5
A-1090 Wien, Austria

ISBN-13:978-3-642-83673-2   e-ISBN-13:978-3-642-83671-8
DOI: 10.1007/978-3-642-83671-8

Library of Congress Cataloging-in-Publication Data. Newton, Roger G. Inverse Schrödinger scattering in three dimensions / R.G. Newton. p. cm. − (Texts and monographs in physics) Bibliography: p. Includes index. ISBN-13:978-3-642-83673-2(U.S.)1.Scattering (Physics)−Mathematics. 2. Inverse problems (Differential equations) 3. Schrödinger equation. I. Title. II. Series. QC20.7.S3N49 1989 530.1′5−dc20 89-11487

© Springer-Verlag Berlin Heidelberg 1989
Softcover reprint of the hardcover 1st edition   1989

The contents was processed by the author using the LaTeX macro package.

Media conversion: Universitätsdruckerei H. Stürtz AG, Würzburg
2155/3150-543210

To Res Jost,
who taught me inverse scattering

Preface

Most of the laws of physics are expressed in the form of differential equations; that is our legacy from Isaac Newton. The customary separation of the laws of nature from contingent boundary or initial conditions, which has become part of our physical intuition, is both based on and expressed in the properties of solutions of differential equations. Within these equations we make a further distinction: that between what in mechanics are called the equations of motion on the one hand and the specific forces and shapes on the other. The latter enter as given functions into the former. In most observations and experiments the "equations of motion," i.e., the structure of the differential equations, are taken for granted and it is the form and the details of the forces that are under investigation.

The method by which we learn what the shapes of objects and the forces between them are when they are too small, too large, too remote, or too inaccessible for direct experimentation, is to observe their detectable effects. The question then is how to infer these properties from observational data. For the theoretical physicist, the calculation of observable consequences from given differential equations with known or assumed forces and shapes or boundary conditions is the standard task of solving a "direct problem." Comparison of the results with experiments confronts the theoretical predictions with nature. To try to infer the unknown forces and shapes from the observational data is the "inverse problem." Isaac Newton himself was the first to solve such a problem when he showed that forces on the planets that do not follow an inverse-square distance dependence would not lead to Kepler's laws.

The inverse scattering and spectral problems in quantum mechanics have a long history that was outlined in some detail in the book by Chadan and Sabatier [CS77]. For many years the only known solutions of these problems were in one spatial dimension (or in three dimensions for central forces, which is effectively a one-dimensional problem). It was only from about 1971 on (after some early attempts by Kay and Moses) that successful assaults on higher–dimensional inverse problems were made, first by Faddeev. And that solution too languished for about fifteen years because of an important unanswered question in it. In the meantime, the known solution of the inverse scattering problem in one dimension led to the quite unexpected and very fruitful development of soliton theory for the solution of nonlinear partial differential equations that describe the propagation of waves in many different physical contexts. Naturally, these results stimulated interest in the question whether similar phenomena exist in higher dimensions. So far these developments have, unfortunately, yielded very limited results (mostly in two space dimensions). As will become clear to the reader of this book, the principal reason for this is the large degree of over-determination in higher-dimensional inverse scattering problems, which prevents us from knowing how to distort a given admissible scattering amplitude in such a way that it remains compatible with the existence of a local potential (except in the trivial case of translation). Thus, in spite of years of efforts by many people, the reader will not find any solitons in this book.

Of course, solitons are not the only reason why inverse scattering problems are of interest. The Schrödinger equation describes, first of all, the motion of particles in quantum mechanics. Thus the inverse scattering problem has important applications for particle scattering in atomic, molecular, and nuclear physics. In these areas interest centers almost exclusively on central forces (or certain specific modifications of them due to the spin of particles). Only in molecular scattering is there a quantum mechanical application of the Schrödinger equation with noncentral forces. The usefulness of the Schrödinger equation, however, is not limited to quantum mechanics. Versions of it arise in the context of acoustic as well as plasma and electromagnetic waves. In many of these contexts the inverse scattering problem has a great deal of practical interest, for important applications in seismology, nondestructive testing, atmospheric profile inversion, and many other fields. The results that have been obtained so far for the Schrödinger equation may not have many direct applications in these areas of practical utility (and in fact they do not at the present time), but the methods that have been successful here are eminently worth studying because they may well work also, suitably modified and adapted, in other contexts. Attempts at such transfers of methodology are presently going on in many directions, and many more are likely to follow. Some of them are characterized by insufficient knowledge of and attention to the mathematical bases on which the known results for the Schrödinger equation rest.

The principal purpose of this book is to present what is known about the three-dimensional inverse scattering problem for the Schrödinger equation without assuming spherical symmetry, including many of the details and proofs. Much of the book's contents has been published in articles in various journals before, but I have made many improvements and modifications of published proofs and conclusions, and a number of new results are given. I will also specifically point out when there are open questions. The book should not be regarded as attempting to summarize a closed area of research. The field is not closed; it may, in fact, be ready for many new developments that carry it closer to applicability than it is now. This book, I hope, will be helpful in that endeavor.

The financial support of the U.S. National Science Foundation for part of the work on which this book is based is gratefully acknowledged. I am also indebted to Professor T. Aktosun for a critical reading of the manuscript.

Bloomington, Indiana   *Roger G. Newton*
June, 1989

Contents

Introduction

Let us assume that we are given a differential equation, or a system of differential equations, that contains certain parameters and a number of arbitrary functions which we shall refer to as the "forces." Also included are boundary conditions of a given kind on arbitrary surfaces. The solution of this system gives rise to a set of functions that are, more or less directly, observable. In many cases these are either connnected with the spectrum of the system of the differential equation *cum* boundary conditions, or with the asymptotic form of solutions at large distances. Let us collectively refer to these functions as the "data". Thus the solution of the direct problem of solving the differential equation or system establishes a map $\mathscr{M}$ from the "forces" to the "data". The inverse problem is to find the inverse $\mathscr{M}^{-1}$ of this map. For reasons that may be partly in the nature of the problem and partly technical, only a certain class of forces may be admitted. For example, in a scattering problem the potential function in the Schrödinger equation has to tend to zero at infinity, and the manner in which it is required to do so is usually dictated by technical considerations. We shall call $\mathscr{P}$ the domain of $\mathscr{M}$; this is the class of admissible forces. Let $\mathscr{D}$ be the set of functions that have the right form and general properties to be considered as possible data. A certain subset $\mathscr{D}^P$ of $\mathscr{D}$ is the image of $\mathscr{P}$ under $\mathscr{M}$. Generally not all conceivable data functions are in fact possible data when the forces are restricted to $\mathscr{P}$.

The first step in solving the inverse problem must clearly be to answer the question whether the map $\mathscr{M}$ is one-to-one. We will take it for granted that $\mathscr{M}$ is well-defined, so that it assigns to each point in $\mathscr{P}$ a unique image. The **uniqueness** problem is the question whether two distinct pre-images may have the same image. If there is no uniqueness, there can obviously be no inverse.

If the one-to-one nature of $\mathscr{M}$ has been established, the inverse map $\mathscr{M}^{-1}$ exists. However, its domain of definition may be quite obscure. The set $\mathscr{D}^P$, i.e., the set of *admissible* data, may be difficult to define in terms other than its definition as the image of $\mathscr{P}$ under $\mathscr{M}$. This is the **characterization** problem, i.e., the problem cf how to characterize admissible data intrinsically, without reference to the set $\mathscr{P}$ and the map $\mathscr{M}$. It may also be called the *existence* problem: For which data in $\mathscr{D}$ does a solution of the inverse problem exist? In general, the smaller $\mathscr{P}$ the more difficult is the characterization problem. It is therefore generally advisable not to restrict the domain of $\mathscr{M}$ more than necessary, even if doing so appears to be innocuous or even desirable from a physical point of view.

The existence problem becomes particularly acute if the data are highly redundant, as they are in the case of the inverse scattering problem in dimensions higher than one. The images under the map $\mathscr{M}$ may be data functions that appear to contain more information than they do; because they are $\mathscr{M}$-images of $\mathscr{P}$, some of that information is, in fact, redundant. In some cases this may be a matter of not being able to specify the precise image of a set $\mathscr{P}$ that is restricted by technical considerations (to make certain proofs easier, say); such instances may give rise to little concern. Other cases may be much more serious. For example, whereas the members of $\mathscr{D}$ may be functions on an n-dimensional manifold, those of $\mathscr{D}^P$ may be restricted to a lower-dimensional submanifold that is difficult to identify. The existence or admissibility problem may then be very difficult. This is, in fact, the situation for the three-dimensional inverse scattering problem.

There is a subclass of inverse problems for which the existence question does not arise. If a certain data function is given and it is known that there is an underlying force in the class $\mathscr{P}$, the aim may be merely to *reconstruct* that force. For example, the data may be synthetic, i.e., constructed from some point in $\mathscr{P}$ by solving the direct problem. Therefore it is useful to distinguish between a *reconstruction problem* and a *construction problem*, in which the existence of an underlying point in $\mathscr{P}$ is not known.

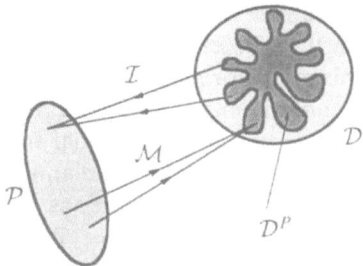

Fig. 1. The mapping $\mathscr{M}$ and its implemented inverse $\mathscr{I}$

If the uniqueness and existence problems have been solved, the next step is to establish the **stability** of the mapping. This is the question of its continuity in some suitable topology. The continuity of the map $\mathscr{M}$ may be well established, but that does not necessarily tell us if the inverse map $\mathscr{M}^{-1}$ is continuous, and if so, in what topology. This question is particularly important here because all experimentally given data have errors. That is to say, even observational data that are known to come from a given force in $\mathscr{P}$, are not exactly equal to the data that would be calculated from that force. The two sets of data will differ by small unkown amounts. If the inversion is not continuous, this small difference in the images may be correlated with a large difference in the pre-images, so that the inferred forces may bear little resemblance to the real ones.

Once existence, uniqueness, and stability have been established, the problem is *well posed* in the sense of Hadamard. For some mathematicians that would be the end of the matter. However, for a physicist the most important task is to find an actual **construction procedure** for the map $\mathscr{M}^{-1}$. Without an algorithm or method of implementing the inversion the solution to the inverse problem would be useless and empty. This will usually involve solving an integral equation or a system of integral equations. The most appealing of the procedures utilize *linear* integral equations. There then arises the need for proving the existence and uniqueness of the solution of such an equation, so that the implementation $\mathscr{I}$ of the map $\mathcal{M}^{-1}$ is well defined. Finally it is necessary to prove that the implemented inverse map $\mathcal{I}$ is one-to-one, in other words, that $\mathcal{I}$ does not map two distinct data functions into the same point in $\mathcal{P}$; again a problem of **uniqueness**. Without such uniqueness there is no assurance that $\mathcal{M}$ will map a given image under $\mathcal{I}$ back into its pre-image, and we do not know that the circle consisting of data $\longrightarrow$ forces $\longrightarrow$ data is closed. For the inverse scattering problem this means that it has to be shown explicitly that the potential constructed by the given procedure from a prescribed scattering amplitude actually leads to that scattering amplitude under $\mathcal{M}$.

Inverse problems are notoriously *ill posed*. Thus it may well happen that one or more of Hadamard's criteria for well-posedness will be found to be violated. (As Pierre Sabatier once observed at a panel discussion, it is the ill-posed problems that are really interesting; once a problem has been proved to be well posed, it is, from a mathematical point of view, essentially solved and hence no longer challenging. For instructive general discussions of inverse problems by Sabatier, see [Sb84 and 87a].) The task then is to find an appropriate modification of the manner in which the problem is formulated so as to make it well posed. This may mean that, in order to achieve uniqueness, additional data will have to be provided. In other cases it may mean that, in order to simplify the characterization or existence problem, the initial set $\mathcal{P}$ may have to be enlarged. The following examples will illustrate my point.

The solution of the inverse scattering problem for the Schrödinger equation in one dimension, with a reflection coefficient given as a function of the wave number, is not unique if there are point eigenvalues (bound states). The same is true in three dimensions for central potentials. In order to make it unique the data have to include the eigenvalues and one additional parameter for each, usually called a "norming constant" (see, for example, [CS77]). This was not known until specific mathematical attention was paid to solving the inverse problem.

In the solution to the inverse scattering problem for the Schrödinger equation in three dimensions at a fixed energy with a central potential the redundancy problem manifests itself as a compatibility problem for scattering amplitudes at different energies. The central potential can be uniquely inferred (within a certain class) from the scattering amplitude at one energy. Therefore two scattering amplitudes at different energies are not necessarily compatible with one another and with the existence of an underlying local potential. No direct solution of this compatibility problem is known. It can be circumvented by enlarging the class $\mathcal{P}$ of admissible potentials to include potentials that depend on the energy. In that case the redundancy or existence problem disappears. There is a similar situation in the inverse scattering problem for central potentials when a single phase shift is given for all energies. Here the compatibility problem of two phase shifts of different angular momenta (to which no direct solution is known either) is finessed by including in $\mathcal{P}$ potentials that depend on the angular momentum.

The remedies that have been used for these very special cases are not always available. They should, however, be kept in mind as possibilities to search for when needed.

The methods that will be studied in detail in this book may be divided, so to speak, horizontally and vertically. By the horizontal division I mean methods based on a Marchenko-procedure, a Gel'fand-Levitan-procedure, or a $\bar{\partial}$-procedure. By a vertical division I am referring to the utilization of the standard scattering-solution of the Schrödinger equation, of a "regular" solution, of a standing-wave solution, or of a class of solutions introduced by Faddeev. Several other methods will be discussed briefly and peripherally for the sake of completeness.

The distinction between a Marchenko-method and a Gel'fand-Levitan-method is sometimes blurred in the literature, even though it is, in principle, clear-cut. The Marchenko-method is based on the relation on the real line between two classes of solutions of the Schrödinger equation, one of which is analytic in the upper half of the complex plane, the other in the lower half, and both have given asymptotics. Thus it is intrinsically a Riemann-Hilbert problem. This was perhaps most clearly stated for the first time by Karlsson in [Ka78] for the case of central potentials. The Gel'fand-Levitan-method, on the other hand, is based on a combination of "triangularity" of the Fourier transform of a class of solutions of the Schrödinger equation with respect to the wave number and the completeness relation for these solutions. The origin of the triangularity in turn is, again, analyticity in $\mathbb{C}$ together with asymptotics. The formal procedure of utilizing these basic ingredients goes back to Kay and Moses in [KM55 and 56]. The $\bar{\partial}$-method is of much more recent vintage and originates from the theory of functions of several complex variables. It was introduced into this field by Nachman and Ablowitz [NA84], Beals and Coifman [BC85, 86, and 87], and Novikov and Henkin [NH86, 87a, and 87b].

The present little volume has a very circumscribed aim. It will deal with the Schrödinger equation in three dimensions only. There can be no question that the ideas it contains are equally applicable to all dimensions greater than one. However, their implementation is not the same in all dimensions. There are, for example, special problems in two dimensions that arise from the fact that the needed Green's function has a logarithmic singularity at the origin. This problem has been solved by Margaret Cheney [Ch82, 84a, 84b, and 85] and I did not want to encumber this book with the complications of that case. To include higher dimensions would have meant cluttering up the notation; furthermore, some of the needed estimates are harder to come by. (In fact, some are not known.) The book does not separately discuss the special case of central potentials. That is treated in detail in the excellent volume by Chadan and Sabatier [CS77], as is the inverse scattering problem in one dimension.

The choice of admissible potentials (i.e., the class $\mathscr{P}$) will be dictated primarily by convenience for the proofs of certain needed theorems. No doubt future workers will be able to enlarge $\mathscr{P}$. On the other hand, our attention will never be restricted to potentials of compact support even though that might facilitate results of apparent interest. The main reason is that significantly more far-reaching results which are obtainable by assuming potentials of compact support are likely to be, in fact, false when that assumption is dropped. Even though from a physical point of view there should be no observable difference between potentials of large but compact support and potentials that decrease rapidly at infinity, such results are then unstable and hence of little use. Potentials that decrease exponentially will, on some rare occasions, be assumed simply because

I was unable to obtain the needed result in any other way, and I am hoping that someone else may.

A word about notation. In keeping with contemporary mathematical usage I am not employing a special typeface or arrows to indicate vectors in $\mathbb{R}^3$. When the sets over which variables range is not clear from the context, they will be explicitly indicated. The complex conjugate of a will be denoted by $\bar{a}$, and its real and imaginary parts will be denoted by $\Re a$ and $\Im a$, respectively. The adjoint of an operator O will be denoted by $O^\dagger$. (Please note that A^* in this book means neither the adjoint of A nor its complex conjugate.) We will often abuse notation by denoting an operator and its integral kernel by the same letter. The unit operator will be denoted by $\mathbb{1}$, the transpose of a kernel g by $\tilde{g}$, the range and the nullspace of the operator O by $\operatorname{ran} O$ and $\operatorname{nul} O$, respectively; if O differs from unity by an operator in the Hilbert-Schmidt class, its modified Fredholm determinant will be denoted by $det_2 O$; $\operatorname{tr} M$ will be the trace of M.

Part I

Use of the Scattering Solution

1. The Direct Scattering Problem

1.1 The Scattering Solution

We wish to solve the Schrödinger equation

$$(\Delta + k^2)\psi = V\psi, \tag{1.1}$$

where Δ is the Laplacian in $\mathbb{R}^3$, with boundary conditions appropriate to scattering. The function V, $\mathbb{R}^3 \mapsto \mathbb{R}$, is not assumed to have any particular symmetry properties but it is assumed to decrease to zero at infinity in a manner to be specified later. The solution is to describe a plane wave sent in the direction of the unit vector θ toward the "scattering center," together with an outgoing spherical wave that describes the response of this center. The time dependence of monochromatic waves being assumed conventionally to be $e^{-i\omega t}$, with $\omega = k^2 > 0$, an outgoing spherical wave is described by e^{ikr}/r, $r = |x|$, and thus the asymptotic boundary condition of the solution of (1.1) sought is

$$\psi^+(k, \theta, x) = e^{ik\theta \cdot x} + (e^{ikr}/r)A(k, \hat{x}, \theta) + o(1/r) \tag{1.2}$$

as $r \to \infty$, where $\hat{x} = x/r$, $k > 0$, $\theta \in \mathbb{R}^3$, $\theta \cdot \theta = 1$. We will regard θ both as a unit vector in $\mathbb{R}^3$ and as a point on the unit sphere S^2. The function A, $\mathbb{R} \times S^2 \times S^2 \mapsto \mathbb{C}$, is the *scattering amplitude*, and its modulus squared is the experimentally observable differential scattering cross section:

$$|A|^2 = d\sigma/d\omega.$$

The direct scattering problem consists of the calculation of A from V.

The boundary condition (1.2) may also be stated in the form of the Sommerfeld *radiation condition*:

$$\psi^+ = e^{ik\theta \cdot x} + \psi_{\mathrm{sc}},$$

$$\lim_{r \to \infty} [k\psi_{\mathrm{sc}} + i\hat{x} \cdot \nabla\psi_{\mathrm{sc}}] = 0,$$

which has the advantage of not containing the unknown function A.

The differential equation (1.1), together with the outgoing-wave boundary condition may be replaced by the integral equation

$$\psi^+(k, \theta, x) = e^{ik\theta \cdot x} + \int_{\mathbb{R}^3} dy\, G_0^+(k, x - y)V(y)\psi^+(k, \theta, y), \tag{1.3}$$

in which $G_0^+(k, x - y)$ is the outgoing-wave Green's function

$$\begin{aligned} G_0^+(k, x) &= \frac{1}{(2\pi)^3} \int_{\mathbb{R}^3} dk' \frac{e^{ik' \cdot x}}{k^2 - k'^2 + i\epsilon} \\ &= -e^{ikr}/4\pi r, \quad k > 0. \end{aligned} \tag{1.4}$$

It is the integral kernel of the boundary value on the real axis of the resolvent $(k^2 + \Delta)^{-1}$ of the self-adjoint extension of the operator $-\Delta$ on $\mathbb{R}^3$, as $\Im k^2 \downarrow 0$. This limit is usually indicated by writing $(k^2 + i\epsilon + \Delta)^{-1}$. The integral equation (1.3) is the *Lippmann-Schwinger equation*. All the properties of ψ^+ that we need are obtained from (1.3).

As it stands, the inhomogeneity of the Lippmann-Schwinger equation is not in $L^2(\mathbb{R}^3)$ and its kernel is not in the Hilbert-Schmidt class. In order to remedy these defects one need only multiply (1.3) by $|V(x)|^{1/2}$ to obtain the integral equation

$$\varphi(k, \theta, x) = \varphi_0(x) + \int_{\mathbb{R}^3} dy\, K_0(k, \theta, x, y)\varphi(k, \theta, y), \tag{1.5}$$

where

$$\varphi_0(x) := |V(x)|^{1/2}, \quad \varphi(k, \theta, x) := |V(x)|^{1/2}\psi^+(k, \theta, x)e^{-ik\theta \cdot x}$$

$$K_0(k, \theta, x, y) := -|V(x)|^{1/2}V^{1/2}(y)e^{ik[|x-y|-\theta\cdot(x-y)]}/4\pi|x-y|, \tag{1.6}$$

$$V^{1/2} := |V|^{1/2}\mathrm{sgn}\, V,$$

and we regard $K_0(k, \theta, x, y)$ as the kernels of the family of operators $K_0(k, \theta)$.

At this point, let us define a certain useful class of potentials.

Definition 1.1.1. $V \in \mathcal{V}$ if $V \in \mathbb{R}$ and $\exists a, C > 0$ such that for all $y \in \mathbb{R}^3$

$$\int_{\mathbb{R}^3} dx\, |V(x)|^2 + \int_{\mathbb{R}^3} dx\, |V(x)| \left(\frac{|x| + |y| + a}{|x - y|}\right)^2 < C.$$

This definition implies that if $V \in \mathcal{V}$ then $V \in L^1 \cap L^2$ and that $\exists b$ such that for all $y \in \mathbb{R}^3$

$$\int_{\mathbb{R}^3} dx\, \frac{|V(x)|}{|x - y|^2} < b.$$

Furthermore, then V is in the *Rollnik class*:

$$\|V\|_R^2 := \int_{\mathbb{R}^3 \times \mathbb{R}^3} dx\, dy\, \frac{|V(x)V(y)|}{|x - y|^2} < \infty. \tag{1.7}$$

Let $\mathcal{V}_0$ be defined by

Definition 1.1.2. $\mathcal{V}_0 = \{V \mid V \in \mathbb{R}, \lim_{|x|\to\infty} V(x) = 0, \text{ and } \exists a, C, \epsilon > 0, \text{ such that for all } x \in \mathbb{R}^3, |\nabla V(x)| < C(a + |x|)^{-4-\epsilon}\}$.

One easily finds that $\mathcal{V}_0 \subset \mathcal{V}$.

Now, if $V \in L^1(\mathbb{R}^3)$ then $\varphi_0 \in L^2$, and if V is in the Rollnik class, then K_0 is Hilbert-Schmidt, $\mathrm{tr}K_0^\dagger K_0 < \infty$, for all $k \in \mathbb{C}^+ \cup \mathbb{R}$ and all $\theta \in \mathbb{R}^3$, $|\theta|^2 = 1$. Therefore (1.5) is a Fredholm equation. Its modified Fredholm determinant will be denoted by Δ. Since that Fredholm determinant is expressible as a power series in which each term is a polynomial in $\mathrm{tr}K_0^n$, one easily sees that

$$\Delta(k) := \det_2(\mathbb{1} - K_0) = \det_2(\mathbb{1} - G_0^+ V),$$

which is independent of θ. Thus the equation (1.5) has a unique solution in L^2 for each nonexceptional value of $k \in \mathbb{C}^+ \cup \mathbb{R}$ and all $\theta \in \mathbb{R}^3$, $|\theta|^2 = 1$. The exceptional points are those values k_0 of k for which $\Delta(k_0) = 0$. For $\Im k > 0$ the kernel K_0 and each term in the series expansions of the Fredholm determinant and of the first Fredholm minor are analytic functions of k, and the two series converge uniformly in any compact subset of $\mathbb{C}^+$. Thus $\Delta(k)$ is holomorphic in $\mathbb{C}^+$ and $\varphi(k, \theta, \cdot)$ is, for each $\theta \in \mathbb{R}^3, |\theta|^2 = 1$, an analytic function of k with values in $L^2(\mathbb{R}^3)$, meromorphic in $\mathbb{C}^+$ with poles at the isolated exceptional points, which are the zeros of $\Delta(k)$. On the real axis $\varphi(k, \theta, \cdot)$ is continuous as a function of k, except at exceptional points, if there are any.

The solution ψ^+ of (1.3) can be recovered from the solution φ of (1.5) by

$$\psi^+(k, \theta, x) = e^{ik\theta \cdot x} \left[1 + \int_{\mathbb{R}^3} dy\, G_0'(k, \theta, x - y) V^{1/2}(y) \varphi(k, \theta, y) \right], \tag{1.8}$$

$$G_0'(k, \theta, x) = -e^{ik(r - \theta \cdot x)}/4\pi r,$$

and since for all nonexceptional $k \in \mathbb{C}^+ \cup \mathbb{R}$

$$\left| \int dy\, G_0' V^{1/2} \varphi \right| \leq \left[\int dy\, \frac{|V(y)|}{|x - y|^2} \right]^{1/2} \|\varphi\|_{L^2} < C$$

if $V \in \mathscr{V}$, for each nonexceptional $k \in \mathbb{C}^+ \cup \mathbb{R}$, and all $\theta \in \mathbb{R}^3$, $|\theta|^2 = 1$, $\psi^+ e^{-ik\theta \cdot x}$ exists and is uniformly bounded for all $x \in \mathbb{R}^3$. Furthermore it is an analytic function of k, meromorphic in $\mathbb{C}^+$ and continuous on $\mathbb{R}$, except possibly at its exceptional points.

Lemma 1.1.3. *Suppose that $V \in \mathscr{V}$ and $\Delta(0) \neq 0$. Then $\exists C$ such that for all $k \in \mathbb{R}$, $\theta, x \in \mathbb{R}^3$, $|\theta| = 1$,*

$$\| [1 - L(k)]^{-1} \| \quad < \quad C, \tag{1.9}$$

$$\left\| \frac{\partial}{\partial k} [1 - L(k)]^{-1} \right\| \quad < \quad C, \tag{1.10}$$

$$|\psi^+(k, \theta, x)| \quad < \quad C, \tag{1.11}$$

$$\left| \frac{\partial}{\partial k} \psi^+(k, \theta, x) \right| \quad < \quad C. \tag{1.12}$$

Here $L(k)$ is the operator family whose integral kernels on $\mathbb{R}^3$ are

$$L(k, x, y) := |V|^{1/2}(x) V^{1/2}(y) G_0^+(k, x - y) \tag{1.13}$$

and $\| \cdot \|$ is the operator norm $L^2(\mathbb{R}^3) \mapsto L^2(\mathbb{R}^3)$. Furthermore, ψ^+ is differentiable with respect to θ.

Proof. The proof of (1.9) follows directly from the absence of exceptional points on the real axis (see the next Section) and (1.98) below. Similarly for (1.10), together with the fact that

$$\frac{d}{dk}(1 - L)^{-1} = (1 - L)^{-1} \frac{dL}{dk}(1 - L)^{-1}, \tag{1.14}$$

where

$$\frac{\partial L}{\partial k}(k, x, y) = -\frac{i}{4\pi}|V|^{1/2}(x)V^{1/2}(y)e^{ik|x-y|}.$$

The pointwise bounds (1.11) and (1.12) follow from (1.9) and (1.10) together with the representation (1.8). The differentiability with respect to θ follows in the same way. $\square$

Since the analytic continuation of G_0^+ via $\mathbb{C}^+$ to the negative real axis leads to

$$G_0^+(-k, x) = \overline{G_0^+(k, x)}$$

it follows from the reality of the potential and from the Lippmann-Schwinger equation that for $k > 0$

$$\psi^+(-k, \theta, x) = \overline{\psi^+(k, \theta, x)}. \tag{1.15}$$

The solution $\psi^+(-k, -\theta, x)$ of (1.1) is usually denoted by ψ^- and called the *incoming-wave solution*:

$$\psi^-(k, \theta, x) := \psi^+(-k, -\theta, x). \tag{1.16}$$

It satisfies (1.3) with G_0^+ replaced by $G_0^- = \overline{G_0^+}$. The scattering solutions are mutually orthogonal and so normalized that

$$\frac{1}{(2\pi)^3} \int_{\mathbb{R}^3} dx\, \overline{\psi^+(k, \theta, x)}\psi^+(k', \theta', x) = \frac{1}{k^2}\delta(k - k')\delta(\theta, \theta'), \tag{1.17}$$

where $\delta(\theta, \theta')$ is the solid-angle Dirac distribution defined so that $\int_{S^2} d\theta\, \delta(\theta, \theta') = 1$ with $d\theta := d\varphi\, d\cos\vartheta$ in spherical polar coordinates.

Suppose that $V(x)$ is replaced by the *translated* potential

$$V_y(x) := V(x + y). \tag{1.18}$$

If V has its "center" at $x = 0$ then V_y has it at $x = -y$. It follows from (1.3) that the corresponding scattering solution ψ_y^+ is given by

$$\psi_y^+(k, \theta, x) = \psi^+(k, \theta, x + y)e^{-ik\theta\cdot y}. \tag{1.19}$$

1.2 Exceptional Points

For $\Im k > 0$ the kernel of the Lippmann-Schwinger equation is Hilbert-Schmidt. If for some $k_0 \in \mathbb{C}^+$ the Fredholm determinant Δ vanishes, $\Delta(k_0) = 0$, then the homogeneous version of (1.3) has a solution, and this solution decreases exponentially at infinity because of the exponential decrease of the Green's function. Thus k_0 is an eigenvalue of (the self-adjoint extension of) the operator

$$H := V - \Delta,$$

i.e., (1.1) has a bound state. Since H is self-adjoint, k_0 must lie on the imaginary axis. Conversely, if k_0^2 is an eigenvalue of H then there must exist an L^2-solution of (1.1) and it must solve the homogeneous version of (1.3); hence $\Delta(k_0) = 0$. Thus there is a one-to-one correspondence between negative eigenvalues $k_0^2 < 0$ and zeros of Δ, $\Delta(k_0) = 0$.

That the total number N_B of negative point eigenvalues of H (counting their multiplicities) is finite if V is in $\mathscr{V}$ and hence in the Rollnik class follows from the Birman-Schwinger bound:

$$N_B \leq \|V\|_R^2/(4\pi)^2. \tag{1.20}$$

where $\| \cdot \|_R$ is the Rollnik norm defined in (1.7). The following is proved in Section 1.7.

Lemma 1.2.1. *If the analytic function $\Delta(k)$ has a zero at $k = k_0$, $\Im k_0 > 0$, the multiplicity of this zero equals the degeneracy of the corresponding bound-state eigenvalue.*

For real k_0 the situation is more complicated. If $\Delta(k_0) = 0$, $\Im k_0 = 0$, then there are two possibilities. Either (i) a given solution of the homogeneous form of (1.5) leads to an L^2-solution of (1.1), or (ii) it does not. If a solution of kind (i) exists, the exceptional point is called *of the first kind* [Ne77a]. If a solution of the kind (ii) exists, the exceptional point is called *of the second kind*. Thus an exceptional point of the first kind implies the existence of a non-negative eigenvalue. It was proved by Kato [Ka59,also La73] that a potential in the class $\mathscr{V}$ produces no bound states of positive eigenvalue. Exceptional points of the second kind for $|k| > 0$ have also been ruled out [As71, Ne77a]. Thus for $V \in \mathscr{V}$ the only zero of $\Delta(k)$ on the real axis can occur at $k = 0$.

If the origin is an exceptional point of the first kind then zero is an L^2-eigenvalue of $V - \Delta$, i.e., there is a bound state of zero energy. If it is an exceptional point of the second kind, there is a so-called *half-bound state*. Both may exist simultaneously, as the origin may be an exceptional point of both the first and the second kind [Ne77a]. The number of linearly independent half-bound solutions is never greater than one, while that of L^2-solutions may be greater than one but must be finite.

Lemma 1.2.2. *Define $L(k)$ as in (1.13). Suppose that $V \in \mathscr{V}$ and that $k = 0$ is an exceptional point. Then Lemma 1.1.3 holds for all $k_0 > 0$ and all $k \geq k_0$. If $k = 0$ is an exceptional point of the first but not of the second kind, then $\exists C \neq 0$ such that as $k \to 0$,*

$$\Delta(k) = Ck^{2N} + o(k^{2N}), \tag{1.21}$$

where N is the dimensionality of the nullspace of $[\mathbb{1} - G_0(0)V]$, and for all $f, g \in L^2(\mathbb{R}^3)$

$$(f, [\mathbb{1} - L(k)]^{-1}g) = k^{-2}(f, |V|^{1/2}\mathscr{P}_b V^{1/2}g) + O(1), \tag{1.22}$$

where $\mathscr{P}_b$ is the orthogonal projection on the nullspace of $[\mathbb{1} - G_0(0)V]$ in $L^2(\mathbb{R}^3)$, which satisfies $\mathscr{P}_b V = 0$ if V is regarded as a vector in $L^2(\mathbb{R}^3)$, and where $(\cdot, \cdot)$ is the inner product on $L^2(\mathbb{R}^3)$. We also have as $k \to 0$

$$\left(f, \frac{d}{dk} \left[k[\mathbb{1} - L(k)]^{-1}\right] g\right) = k^{-2}(f, |V|^{1/2}\mathscr{P}_b \Gamma_1 \mathscr{P}_b V^{1/2}g)$$

$$+ k^{-1}(f, |V|^{1/2}\mathscr{P}_b \Gamma_2 \mathscr{P}_b V^{1/2}g) + O(1), \tag{1.23}$$

where $\Gamma_1(x, y) := \delta(x - y) - |x - y|$, and $\Gamma_2(x, y) := -i|x - y|^2$.

If $k = 0$ is an exceptional point of the second and not of the first kind, then $\exists C \neq 0$ such that as $k \to 0$

$$\Delta(k) = Ck + o(k), \tag{1.24}$$

and for all $f, g \in L^2(\mathbb{R}^3)$

$$(f, [\mathbb{1} - L(k)]^{-1}g) = \frac{4\pi}{ik} \int_{\mathbb{R}^3} dx\, \overline{f} |V|^{1/2} u \int_{\mathbb{R}^3} dy\, u V^{1/2} g + O(1), \tag{1.25}$$

where $G_0(0)Vu = u$, $u \in \mathbb{R}$, and $(\int dx V u)^2 = 1$. We also have as $k \to 0$

$$\left(f, \frac{d}{dk} \left[k[\mathbb{1} - L(k)]^{-1} \right] g \right) = O(1). \tag{1.26}$$

If $k = 0$ is an exceptional point of both kinds, then a combination of the above terms occurs.

This lemma will be proved in Section 1.7.

Let us now consider the bound-state eigenfunctions of (1.1) with negative eigenvalues. Suppose that $-\kappa^2 < 0$ is an eigenvalue of multiplicity N and that $\{u_\kappa^a(x), a = 1,\ldots,N\}$ is an arbitrary set of real, linearly independent eigenfunctions,

$$\int_{\mathbb{R}^3} dx\, u_\kappa^a(x) u_\kappa^b(x) = e_{ab}^\kappa. \tag{1.27}$$

[Since V is real, the eigenfunctions can be chosen real too.] The $N \times N$ matrix e_κ whose entries are e_{ab}^κ is real, symmetric and invertible. We shall denote its inverse by d_κ. Each function $|V|^{1/2}u_\kappa^a$ is at the same time an eigenfunction of the compact operator whose kernel is $K_0(i\kappa, \theta, x, y)$, with the eigenvalue 1. Therefore N is necessarily finite [which, of course also follows from (1.20)].

The asymptotic form of u for $r = |x| \to \infty$ is obtained from the homogeneous form of the Lippmann-Schwinger equation:

$$u_\kappa^a(x) = -(e^{-\kappa r}/4\pi r) Y_\kappa^a(\hat{x}) + o(e^{-\kappa r}/r) \tag{1.28}$$

where $\hat{x} = x/|x|$ and the *characters* Y_κ^a are given by

$$Y_\kappa^a(\theta) := \int_{\mathbb{R}^3} dx\, V(x) e^{\kappa \theta \cdot x} u_\kappa^a(x), \tag{1.29}$$

where $\theta \in \mathbb{R}^3$, $\theta \cdot \theta = 1$. They uniquely characterize the eigenfunctions u_κ^a; any two of them are linearly independent functions, $S^2 \mapsto \mathbb{R}$, if and only if the corresponding functions u are linearly independent. (For $\kappa \neq \kappa'$, of course, the characters are not necessarily linearly independent.) If the eigenfunctions u_κ^a are chosen real then so are the character functions.

Definition 1.2.3. We denote the linear span of the character functions $Y_\kappa^a(\theta)$, $a = 1,\ldots,N$, by $\mathscr{H}_\kappa$ and define Q_κ, with the integral kernel $Q_\kappa(\theta, \theta') = Q_\kappa(\theta', \theta) = \overline{Q}_\kappa(\theta, \theta')$, to be the orthogonal projection on $\mathscr{H}_\kappa$.

It is of later interest to determine the asymptotic form of the analytic continuation of the scattering solution ψ^+ for large $|x|$. For this purpose we express ψ^+ in terms of the kernel G of the resolvent $(k^2 - H)^{-1}$, whose boundary value is G^+, the *complete* Green's function:

$$\psi^+(k, \theta, x) = e^{ik\theta \cdot x} + \int_{\mathbb{R}^3} dy \, G^+(k, x, y) V(y) e^{ik\theta \cdot y}. \tag{1.30}$$

If $-\kappa^2$ is an eigenvalue of H, the principal part of G at its pole (which is necessarily simple because H is self-adjoint if V is real) at $k = i\kappa$ is given by

$$G(k, x, y) = \frac{\sum_{a,b=1}^N u_\kappa^a(x) d_{ab}^\kappa u_\kappa^b(y)}{2i\kappa(k - i\kappa)} + \cdots,$$

from which it follows that the analytic continuation of ψ^+ has a simple pole there and its principal part is

$$\psi(k, \theta, x) = \frac{1}{2i\kappa(k - i\kappa)} \sum_{a,b=1}^N Y_\kappa^a(-\theta) d_{ab}^\kappa u_\kappa^b(x) + \cdots. \tag{1.31}$$

Thus the asymptotic form of the residue of ψ at $k = i\kappa$ for large r is

$$\mathrm{Res}_{i\kappa}\psi(k, \theta, x) = \frac{i}{8\pi\kappa} \frac{e^{-\kappa r}}{r} \sum_{a,b} Y_\kappa^a(-\theta) d_{ab}^\kappa Y_\kappa^b(\hat{x}) + o\left(\frac{e^{-\kappa r}}{r}\right), \tag{1.32}$$

where $\hat{x} = x/|x|$.

The particular choice of eigenfunctions $u_\kappa^a(x)$ to span the eigenspace of H at the eigenvalue κ may be avoided by using the orthogonal projection P_κ onto it, $P_\kappa^\dagger = P_\kappa$, $P_\kappa^2 = P_\kappa$. Its integral kernel is given by

$$P_\kappa(x, y) = \sum_{a,b} u_\kappa^a(x) d_{ab}^\kappa u_\kappa^b(y), \tag{1.33}$$

so that $\mathrm{tr} P_\kappa = N$. The asymptotic form of $P_\kappa(x, y)$ for large $r = |x|$ is

$$P_\kappa(x, y) = -(e^{-\kappa r}/4\pi r) P_\kappa^\infty(\hat{x}, y) + o(e^{-\kappa r}/r),$$

where $\hat{x} = x/|x|$ and

$$\begin{aligned} P_\kappa^\infty(\theta, y) &= \sum_{a,b} Y_\kappa^a(\theta) d_{ab}^\kappa u_\kappa^b(y) \\ &= \int_{\mathbb{R}^3} dx \, V(x) e^{-\kappa\theta \cdot x} P_\kappa(x, y). \end{aligned} \tag{1.34}$$

The residue of $G(k, x, y)$ at $k = i\kappa$ can now be expressed as $P_\kappa(x, y)/2i\kappa$ and that of $\psi(k, \theta, x)$,

$$\mathrm{Res}_{i\kappa}\psi(k, \theta, x) = \frac{1}{2i\kappa} P_\kappa^\infty(-\theta, x). \tag{1.35}$$

The asymptotic form of this residue for large $|x| = r$ is

$$\mathrm{Res}_{i\kappa}\psi(k, \theta, x) = \frac{i}{8\pi\kappa} \frac{e^{-\kappa r}}{r} Y_\kappa(-\theta, \hat{x}) + o\left(\frac{e^{-\kappa r}}{r}\right), \tag{1.36}$$

where

$$Y_\kappa(\theta, \theta') = \sum_{a,b} Y_\kappa^a(\theta) d_{ab}^\kappa Y_\kappa^b(\theta')$$

$$= \int_{\mathbb{R}^3 \times \mathbb{R}^3} dx\, dy\, V(x) V(y) P_\kappa(x, y) e^{\kappa(\theta \cdot x + \theta' \cdot y)}. \tag{1.37}$$

Now define derivatives of the characters by means of the differential operator

$$D_\xi(\theta) := \frac{1}{|\theta||\xi|} (\theta \xi \nabla_\theta) \tag{1.38}$$

where (abc) denotes the triple vector product and ∇_θ is the gradient with respect to θ. Here $\xi \in \mathbb{R}^3$, $|\xi|^2 = 1$, is a unit vector in the direction of rotation. If D_ξ acts on a function of $a \cdot \theta$, $f(a \cdot \theta)$, then $D_\xi f = (a\theta\xi)f'$, where f' is the derivative of f. Then we define

$$Y_{\kappa\xi}^{a(m)}(\theta) := D_\xi^m Y_\kappa^a(\theta). \tag{1.39}$$

Thus by (1.29)

$$Y_{\kappa\xi}^{a(m)}(\theta) = \kappa^m \int_{\mathbb{R}^3} dx\, V(x)(x\theta\xi)^m u_\kappa^a(x) e^{\kappa\theta \cdot x}. \tag{1.40}$$

It is clear from this that for the existence of $Y_{\kappa\xi}^{a(m)}$ we have to require that $|x|^m V \in L^1(\mathbb{R}^3)$. We furthermore define

$$Y_{\kappa\xi}^{mn}(\theta, \theta') := \sum_{a,b=1}^N Y_{\kappa\xi}^{a(m)}(\theta) d_{ab}^\kappa Y_{\kappa\xi}^{b(n)}(\theta'). \tag{1.41}$$

Let us define $Y_{\kappa\xi}^M$ to be the $(M+1) \times (M+1)$ matrix whose entries are $Y_{\kappa\xi}^{mn}(\theta, -\theta)$, and

$$y_{\kappa\xi}^M := \det Y_{\kappa\xi}^M. \tag{1.42}$$

Regarding $Y_{\kappa\xi}^{a(n)}(\theta)$, $a = 1, \ldots, N$, $n = 0, \ldots, M$, for fixed θ and ξ, as the N components of $M+1$ vectors in $\mathbb{R}^N$, we conclude that if $M \geq N$ there must exist a linear relation between them:

$$\sum_{n=0}^M c_{\kappa\xi}^n(\theta) Y_{\kappa\xi}^{a(n)}(\theta) = 0, \quad a = 1, \ldots, N, \tag{1.43}$$

where not all $c_{\kappa\xi}^n$ vanish. It then follows from (1.41) that

$$\sum_{n=0}^M Y_{\kappa\xi}^{mn}(\theta, \theta') c_{\kappa\xi}^n(\theta') = 0, \quad m = 0, \ldots, M. \tag{1.44}$$

Consequently we have the following result for the determinant $y_{\kappa\xi}^M$.

Lemma 1.2.4. *Assume that* $(1 + |x|^M)V \in L^1$ *and* $-\kappa^2$ *is a bound-state eigenvalue of* $H = V - \Delta$ *with the multiplicity* N. *Then for all* $M \geq N$ *and all* $\theta, \xi \in S^2$,

$$y_{\kappa\xi}^M(\theta) = 0.$$

Thus, the largest M for which $y_{\kappa\xi}^M \not\equiv 0$ is strictly less than the degeneracy.

Definition 1.2.5. A bound state eigenvalue will be called *normal* if for all $\xi \in S^2$ the largest M for which $y_{\kappa\xi}^M(\theta) \not\equiv 0$ is equal to $N-1$, and the nullspace of the $(N+1) \times (N+1)$ matrix $Y_{\kappa\xi}^N(\theta)$ is one-dimensional. Here N is the dimension of the eigenspace of $H = V - \Delta$ at the eigenvalue $-\kappa^2$, $Y_{\kappa\xi}^M$ is defined above in terms of the characters (1.28) and their derivatives (1.39), and $y_{\kappa\xi}^M$ is the determinant of $Y_{\kappa\xi}^M$. A potential $V(x)$ will be called *normal* if all the eigenvalues of $H = V - \Delta$ are normal.

Corollary 1.2.6. *If an eigenvalue is normal then the smallest integer M' such that (1.43) holds with nontrivial $c_{\kappa\xi}^n$ for all $M \ge M'$ equals N.*

This immediately follows from Lemma 1.2.4 because if the smallest such M' were less than N then for that M' (1.44) would also follow and hence the eigenvalue would not be normal.   □

1.3 Completeness

Under various conditions that are fulfilled if $V \in \mathcal{V}_0$, the essential spectrum of H is known to be absolutely continuous [RS79, pp. 439 and 448]. The scattering solutions ψ^+, together with the bound-state eigenfunctions, then span the Hilbert space L^2 in the sense of generalized Fourier-integrals [RS79, p. 99]. We shall write the so-called *completeness relation* in the customary way

$$\frac{1}{(2\pi)^3} \int_0^\infty dk\, k^2 \int_{S^2} d\theta\, \psi^+(k,\theta,x)\overline{\psi^+(k,\theta,y)} + \sum_{\kappa_n} \sum_{a,b=1}^{N_n} u_{\kappa_n}^a(x) d_{ab}^{\kappa_n} u_{\kappa_n}^b(y)$$

$$= \delta(x-y), \tag{1.45}$$

which is a short-hand for the eigenfunction expansion for all $f \in L^2(\mathbb{R}^3)$

$$f(x) = \frac{1}{(2\pi)^3} \int_0^\infty dk\, k^2 \int_{S^2} d\theta\, \psi^+(k,\theta,x) g(k,\theta) + \sum_{\kappa_n} \sum_{a=1}^{N_n} u_{\kappa_n}^a(x) g_n^a, \tag{1.46}$$

$$g(k,\theta) = \int_{\mathbb{R}^3} dx\, f(x)\, \overline{\psi^+(k,\theta,x)},$$

$$g_n^a = \int_{\mathbb{R}^3} dx\, f(x) \sum_b d_{ab}^{\kappa_n} u_{\kappa_n}^b(x).$$

The integrals are all meant in the sense of limits in the mean, and we have Parseval's relation

$$\int_{\mathbb{R}^3} dx\, |f(x)|^2 = \frac{1}{(2\pi)^3} \int_0^\infty dk\, k^2 \int_{S^2} |g(k,\theta)|^2 + \sum_{\kappa_n} \sum_{a,b=1}^{N_n} g_n^a d_{ab}^{\kappa_n} \overline{g}_n^b. \tag{1.47}$$

The completeness relation (1.45) implies the following expansions for the Green's functions G^+ and $G^- = G^{+\dagger}$:

$$G^{\pm}(k, x, y) = \frac{1}{(2\pi)^3} \int_0^{\infty} dk' k'^2 \int_{S^2} d\theta \, \frac{\psi^+(k', \theta, x)\overline{\psi^+(k', \theta, y)}}{k^2 - k'^2 \pm i\epsilon}$$

$$+ \sum_{\kappa_n} \sum_{a,b=1}^{N_n} \frac{u_{\kappa_n}^a(x) d_{ab}^{\kappa_n} u_{\kappa_n}^b(y)}{k^2 + \kappa_n^2}, \tag{1.48}$$

from which it follows that

$$G^+(k, x, y) - G^-(k, x, y) := -2\pi i \delta (k^2 - H)$$

$$= \frac{-ik}{8\pi^2} \int_{S^2} d\theta \, \psi^+(k, \theta, x)\overline{\psi^+(k, \theta, y)}. \tag{1.49}$$

The Dirac distribution here may be regarded simply as a symbolic representation of the right-hand side.

A more general set of expansion theorems than (1.45) was recently proved by Rose and Cheney [RC88] on the assumption that V is real, locally in L^2 and decays at infinity like $|x|^{-5/2-\epsilon}$ for some $\epsilon > 0$. Define the functions

$$\begin{aligned}
v(k, \theta, \theta', x) &:= e^{-ik\theta \cdot x}\psi^+(k, \theta', x), \\
v'(k, \theta, \theta', x) &:= -ik\theta \cdot \nabla v(k, \theta, \theta', x), \\
v_n^b(\theta, x) &:= e^{\kappa_n \theta \cdot x} u_n^b(x), \\
v_n^{b'}(\theta, x) &:= \kappa_n \theta \cdot \nabla v_n^b(\theta, x),
\end{aligned} \tag{1.50}$$

in terms of the scattering solutions and the orthonormal eigenfunctions. Then

$$f(x) = \frac{1}{(2\pi)^3} \int_{-\infty}^{\infty} dk \int_{S^2} d\theta' v'(k, \theta, \theta', x) g(k, \theta, \theta')$$

$$+ \sum_{n,b} \frac{1}{\kappa_n^2} [v_n^{b'}(\theta, x) g_n^b(\theta) + v_n^{b'}(-\theta, x) g_n^b(-\theta)], \tag{1.51}$$

where

$$g(k, \theta, \theta') = -\frac{1}{4\pi} \int_{\mathbb{R}^3} dx f(x)\bar{v}(k, \theta, \theta' x),$$

$$g_n^b(\theta) = \int_{\mathbb{R}^3} dx f(x) v_n^b(-\theta, x),$$

again in the sense of limits in the mean. It should be noted that in this expansion the direction of the unit vector θ is entirely arbitrary. The functions v', however, have not been proved to be linearly independent. Therefore the coefficients g in (1.51) are not known to be unique.

1.4 Asymptotics for Large $|k|$

The behavior of the scattering solution for large $|k|$, both on the real axis and in $\mathbb{C}^+$ is going to play an important role in the further development. This will be obtained from the integral equation (1.5) with the kernel K_0 defined by (1.6). We will, however, have to restrict the potential to a smaller class than $\mathcal{V}$.

Definition 1.4.1. $V \in \mathcal{W}$ if the following conditions are satisfied:

(i) $V \in \mathcal{V}$;

(ii) $\exists C$, $s > 1$, such that for all $x \in \mathbb{R}^3$, $|V(x)|(1 + |x|^2)^s \le C$;

(iii) $\exists C$ such that for all $x \in \mathbb{R}^3$

$$\int_{\mathbb{R}^3} dy \frac{|\nabla V(y)|}{|x - y|} \le C;$$

(iv) $\exists \epsilon$, $\frac{1}{2} < \epsilon < \frac{3}{4}$ such that for each $y \in \mathbb{R}^3$ there exists a constant C and a function $N(t)$, $\mathbb{R}_+ \mapsto \mathbb{R}_+$,

$$\int_0^1 dt\, t^{3/2} N(t) < \infty,$$

$$F^2(s) := \int_s^\infty dt\, t N(t) < C s^{-2\epsilon},$$

and for all $x \in \mathbb{R}^3$, $|\nabla V(x + y)| \le N(|x|)$.

If $V \in \mathcal{V}_0$ then $V \in \mathcal{W}$.

Both for the Fredholm determinant of the Lippmann-Schwinger equation and for the scattering solution we have to distinguish between large-$|k|$ behavior on the real axis and in $\mathbb{C}^+$. On the real axis we need only certain integrability conditions that ensure the existence of a Fourier transform with useful properties. It is clear, on the other hand, that for $k \in \mathbb{C}^+$, ψ^+ must increase exponentially when $x \cdot \theta < 0$. It is therefore necessary to take out the factor $e^{ik\theta \cdot x}$ and to define

$$\zeta(k, \theta, x) := \psi^+(k, \theta, x) e^{-ik\theta \cdot x}. \tag{1.52}$$

This function satisfies the modified Schrödinger equation

$$(\Delta + 2ik\theta \cdot \nabla)\zeta = V\zeta. \tag{1.53}$$

Of course, it has an analytic continuation into $\mathbb{C}^+$ with similar properties as ψ^+; that is, it is meromorphic there with simple poles on the imaginary axis at those points $i\kappa$ for which $-\kappa^2$ is a point eigenvalue.

The large-$|k|$ behavior of ζ and of the Fredholm determinant $\Delta(k)$ are given by the following.

Lemma 1.4.2. *Assume that* $V \in \mathcal{W}$. *Then for* $k \in \mathbb{C}^+$ *and for each* $x \in \mathbb{R}^3$ *and* $\theta \in \mathbb{S}^2$

$$\lim_{|k| \to \infty} \zeta(k, \theta, x) = 1, \tag{1.54}$$

and for $k \in \mathbb{C}^+ \cup \mathbb{R}$

$$\lim_{|k| \to \infty} \Delta(k) = 1. \tag{1.55}$$

Furthermore; if $k = 0$ *is not an exceptional point then*

$$\int_{-\infty}^\infty dk\, |\zeta(k, \theta, x) - 1|^2 < \infty. \tag{1.56}$$

[If $k = 0$ *is an exceptional point then the same holds for any integral that has an interval around the origin removed.]*

The proof of this lemma will be given in Section 1.7.

1.5 Scattering Amplitude and S Matrix

Let $\varphi(x)$ be any uniformly bounded solution of (1.1) for some $k^2 > 0$. Such a solution must have the asymptotic form for large $r = |x|$

$$\varphi(x) = h_1(\hat{x}) \frac{e^{-ikr}}{r} - h_2(\hat{x}) \frac{e^{ikr}}{r} + o\left(\frac{1}{r}\right), \tag{1.57}$$

where $\hat{x} = x/|x|$ and h_1 and h_2 are distributions that contain nothing worse than Dirac deltas. The asymptotic form (1.2) of ψ^+ can be written in this language as

$$\psi^+(k, \theta, x) = \frac{2\pi i}{kr} \left\{ e^{-ikr} \delta(\theta, -\hat{x}) - e^{ikr} \left[\delta(\hat{x}, \theta) - \frac{k}{2\pi i} A(k, \hat{x}, \theta) \right] \right\} + o\left(\frac{1}{r}\right) \tag{1.58}$$

because of the well-known formula for $r = |x| \to \infty$

$$e^{ik\theta \cdot x} = \frac{2\pi i}{kr} [e^{-ikr} \delta(\theta, -\hat{x}) - e^{ikr} \delta(\theta, \hat{x})] + o\left(\frac{1}{r}\right). \tag{1.59}$$

On the other hand, the function

$$\varphi'(x) := \frac{k}{2\pi i} \int_{S^2} d\theta \, h_1(-\theta) \psi^+(k, \theta, x)$$

is a bounded solution of (1.1) which by (1.2) has the asymptotic form

$$\varphi'(x) = \frac{e^{-ikr}}{r} h_1(\hat{x}) - \frac{e^{ikr}}{r} h_3(\hat{x}) + o\left(\frac{1}{r}\right),$$

where

$$h_3(\theta) = h_1(-\theta) - \frac{k}{2\pi i} \int_{S^2} d\theta' \, A(k, \theta, \theta') h_1(-\theta').$$

It follows that $\varphi' = \varphi$, because otherwise the difference $\varphi' - \varphi$ would be a nontrivial solution without any incoming waves. Such a solution is well known not to exist. Therefore we have

Lemma 1.5.1. *The scattering solutions $\psi^+(k, \theta, x)$, with $k \in \mathbb{R}$ fixed and θ ranging over S^2, span the space of bounded solutions of (1.1) in the sense that if $\varphi(x)$ is a uniformly bounded solution of (1.1) then it can be expressed as an integral*

$$\varphi(x) = \int_{S^2} d\theta \, h(\theta) \psi^+(k, \theta, x),$$

where h is a distribution that contains nothing worse than Dirac deltas. The asymptotic form of $\varphi(x)$ for large $r = |x|$ is

$$\varphi(x) = \frac{2\pi i}{k} \left[h(-\hat{x}) \frac{e^{-ikr}}{r} - h'(\hat{x}) \frac{e^{ikr}}{r} \right] + o\left(\frac{1}{r}\right)$$

where $\hat{x} = x/|x|$ and

$$h'(\theta) = h(\theta) - \frac{k}{2\pi i} \int_{S^2} d\theta' \, A(k, \theta, \theta') h(\theta').$$

The distribution kernel

$$S(k, \theta, \theta') := \delta(\theta, \theta') - \frac{k}{2\pi i} A(k, \theta, \theta') \tag{1.60}$$

is called the S matrix. Thus we have the

Corollary 1.5.2. *The amplitude $-h'(\hat{x})$ of the asymptotically outgoing wave of any bounded solution of (1.1) is related to the amplitude $h(-\hat{x})$ of its asymptotically incoming wave by*

$$h'(\theta) = \int_{S^2} d\theta' \, S(k, \theta, \theta') h(\theta')$$

where $S(k, \theta, \theta')$ is given by (1.60) in terms of the scattering amplitude defined by (1.2).

Remark. This corollary shows that the scattering amplitude, or the S matrix, may be defined by any sufficiently large family of bounded solutions of (1.1) in place of ψ^+. All such solutions lead to the same S matrix.

The function $\psi^-(k, \theta, x) = \psi^+(-k, -\theta, x)$ is certainly one of the bounded solutions of (1.1). By (1.2) and (1.60) the function $h(\hat{x})$ for $\psi^+(-k, -\theta, x)$ is $h(\hat{x}) = S(-k, -\hat{x}, -\theta)$. Lemma 1.5.1 therefore implies that

$$\psi^+(-k, \theta, x) = \int_{S^2} d\theta' S(-k, -\theta', \theta) \psi^+(k, \theta', x). \tag{1.61}$$

The scattering amplitude is defined by the asymptotic form (1.2) of ψ^+ for large $|x|$. It follows from the Lippmann-Schwinger equation and the specific form (1.4) of the outgoing-wave Green's function that it has the representation

$$A(k, \theta, \theta') = -\frac{1}{4\pi} \int_{\mathbb{R}^3} dx \, V(x) e^{-ik\theta \cdot x} \psi^+(k, \theta', x) \tag{1.62}$$

$$= -\frac{1}{4\pi} \int_{\mathbb{R}^3 \times \mathbb{R}^3} dx dy \, e^{ik(\theta' \cdot y - \theta \cdot x)} V^{\frac{1}{2}}(x) |V|^{\frac{1}{2}}(y)$$

$$\times \left[[\mathbb{1} - L(k)]^{-1} \right](x, y), \tag{1.63}$$

where L is defined by (1.13). The following properties of A then follow directly from the continuity of ψ^+ and Lemmas 1.1.3 and 1.2.2:

Lemma 1.5.3. *If $V \in \mathscr{V}$ then for each fixed $\theta, \theta' \in \mathbb{R}^3$, $|\theta| = |\theta'| = 1$ the function $A(k, \theta, \theta')$, $\mathbb{R} \mapsto \mathbb{C}$, is continuous and uniformly bounded; furthermore $\exists C$ such that for all $k \in \mathbb{R}$*

$$\left| \frac{\partial}{\partial k} \left[\frac{|k|}{1 + |k|} A(k, \theta, \theta') \right] \right| < C.$$

What is more, $A(k, \theta, \theta')$ is differentiable with respect to θ and θ'.

The scattering amplitude A will, in general, not be the boundary value of an analytic function of k. However, let us consider the *forward scattering amplitude*

$$A(k, \theta, \theta) = -\frac{1}{4\pi} \int_{\mathbb{R}^3} dx \, V(x) \zeta(k, \theta, x) \tag{1.64}$$

where ζ is defined by (1.52). Since for $V \in L^1$, ζ and its k-derivatives are bounded uniformly in k and x for all $x \in \mathbb{R}^3$ and k in any compact subset of $\mathbb{C}^+$ that contains no exceptional points, $A(k, \theta, \theta)$ is, for each $\theta \in \mathbb{S}^2$, an analytic function of k meromorphic in $\mathbb{C}^+$ with simple poles at the points $k = i\kappa_n$ if $-\kappa_n^2$ is a bound state. This is usually referred to as *forward analyticity*. The residues of $A(k, \theta, \theta)$ at these poles can be obtained from (1.35) and (1.34), namely,

$$\text{Res}_{i\kappa} A(k, \theta, \theta) = -\frac{1}{8\pi i\kappa} Y_\kappa(\theta, -\theta), \tag{1.65}$$

where Y_κ is defined by (1.37). Let the degeneracy of the bound state at $-\kappa^2$ be N.

There are considerably more detailed connections between the forward scattering amplitude and the characters of the bound states than is indicated by equation (1.65) and not all of them have been fully explored.

The Definition (1.41) allows us to express the various angle derivatives of the residues of the forward amplitude:

$$A_{\kappa\xi}^{mn}(\theta) := \mathsf{D}_\xi^m(\theta)\mathsf{D}_\xi^n(\theta') \, \text{Res}_{i\kappa} A(k, \theta, \theta')\Big|_{\theta=\theta'} = \frac{i}{8\pi\kappa} Y_{\kappa\xi}^{mn}(\theta, -\theta). \tag{1.66}$$

Suppose now we allow m and n to run from 0 to M and define $A_{\kappa\xi}^M$ to be the $(M+1) \times (M+1)$ matrix whose entries are $A_{\kappa\xi}^{mn}(\theta)$, $m, n = 0, \ldots, M$, and

$$a_{\kappa\xi}^M := \det A_{\kappa\xi}^M. \tag{1.67}$$

Equation (1.66) then says that $a_{\kappa\xi}^M$ differs from $y_{\kappa\xi}^M$, defined by (1.42), by a constant factor. Furthermore, (1.66) implies that equation (1.44) is equivalent to

$$\sum_{n=0}^{M} A_{\kappa\xi}^{nm}(\theta) c_{\kappa\xi}^n(-\theta) = 0, \quad m = 0, \ldots, M. \tag{1.68}$$

Definition 1.5.4. Let $i\kappa$ be a point at which the forward scattering amplitude $A(k, \theta, \theta)$ has a pole and let N' be the smallest integer M_0 such that for all $M \geq M_0$, $a_{\kappa\xi}^M \equiv 0$, where $a_{\kappa\xi}^M$ is defined by (1.67) and (1.66) in terms of the derivatives of the scattering amplitude. The pole will be called *normal* if the nullspace of $A_{\kappa\xi}^{N'}$ is one-dimensional. A scattering amplitude will be called *normal* if all the poles of the forward amplitude are normal.

The following then are immediate consequences of (1.66).

Lemma 1.5.5. *If a pole of the forward scattering amplitude is normal then the degeneracy N of the corresponding eigenvalue is $N \geq N'$.*

Lemma 1.5.6. *If an eigenvalue is normal then the corresponding pole of the forward scattering amplitude is normal and $N' = N$. If the potential is normal then the associated scattering amplitude is normal.*

Furthermore there is the following direct link between the characters and the matrix $A_{\kappa\xi}^M$ for normal eigenvalues.

Theorem 1.5.7. *Assume that $(1 + |x|^M)V \in L^1$ and $-\kappa^2$ is a bound-state eigenvalue of $H = V - \Delta$ with the multiplicity N. If κ is normal then there exists a unique (to within a common factor) set of $N + 1$ functions $c_{\kappa\xi}^m(\theta)$, $m = 0, \ldots, N$ such that the characters $Y_{\kappa\xi}^a$, $a = 1, \ldots, N$ are N linearly independent solutions of the ordinary linear differential equation (1.43) of order N (for fixed ξ). The functions $c_{\kappa\xi}^m(\theta)$ are the components of the ray that is the nullspace of the transpose of $A_{\kappa\xi}^N(-\theta)$. By choosing two linearly independent directions ξ it is therefore possible to determine a complete set of N character functions and thus to determine their span $\mathcal{H}_\kappa$ by means of the equation (1.43).*

The Lippmann-Schwinger equation allows us to express the scattering amplitude in the alternative form

$$A(k, \theta, \theta') = -\frac{1}{4\pi} \int_{\mathbb{R}^3} dx\, V(x)\overline{\psi^-(k, \theta, x)}e^{ik\theta' \cdot x}, \tag{1.69}$$

where ψ^- is defined by (1.16). It therefore follows from (1.16) and (1.15) that A has the symmetry property

$$A(k, \theta, \theta') = A(k, -\theta', -\theta), \tag{1.70}$$

which is called *reciprocity*. In addition, the symmetry (1.15) and the reality of the potential imply that

$$A(-k, \theta, \theta') = \overline{A(k, \theta, \theta')}. \tag{1.71}$$

The S matrix was defined by (1.60) in terms of A. According to Lemma 1.5.1 any bounded solution of (1.1) has the asymptotic form

$$\varphi(x) = \frac{2\pi i}{kr} \left[h(-\hat{x})e^{-ikr} - h'(\hat{x})e^{ikr} \right] + o\left(\frac{1}{r}\right),$$

where $\hat{x} = x/|x|$ and

$$h'(\theta) = \int_{S^2} d\theta'\, S(k, \theta, \theta')h(\theta').$$

The same argument applied to $\overline{\varphi(x)}$ implies that

$$\overline{h(\theta)} = \int_{S^2} d\theta'\, S(k, -\theta, -\theta')\overline{h'(\theta')},$$

which leads to

$$h'(\theta) = \int_{S^2} d\theta' \left[\int_{S^2} d\theta'' S(k, \theta, \theta'')\overline{S(k, -\theta'', -\theta')} \right] h'(\theta'),$$

as well as

$$h(\theta) = \int_{S^2} d\theta' \left[\int_{S^2} d\theta'' \, \overline{S(k, -\theta, -\theta'')}S(k, \theta'', \theta') \right] h(\theta').$$

Since either h or h' may be arbitrarily chosen it follows by means of (1.70) that $S(k, \theta, \theta')$ is the distribution kernel of a unitary operator $S(k)$:

$$S(k)S^\dagger(k) = S^\dagger(k)S(k) = \mathbb{1}. \tag{1.72}$$

It will be very useful to consider, similarly, $A(k, \theta, \theta')$ with $\theta, \theta' \in S^2$, as the kernel of an operator-valued function of $k \in \mathbb{R}$. With a slight abuse of notation we will denote these operators and their kernels by the same letters A and S, respectively. In that notation we write (1.60) in the form

$$S(k) = \mathbb{1} - \frac{k}{2\pi i} A(k). \tag{1.73}$$

Since for each fixed $k \in \mathbb{R}$, $A(k, \theta, \theta')$ is uniformly bounded, the operator A is Hilbert-Schmidt and has a finite trace, and the Fredholm determinant of S is well defined.

Lemma 1.5.8.

$$\det S(k) = \frac{\overline{\Delta(k)}}{\Delta(k)} e^{-\frac{ik}{2\pi}\langle V \rangle}, \tag{1.74}$$

where Δ is the modified Fredholm determinant of the Lippmann-Schwinger equation and $\langle V \rangle := \int_{\mathbb{R}^3} dx \, V(x)$.

This is proved in Section 1.7.

Because S is unitary we may define, mod π, a real number δ by

$$\det S = e^{2i\delta}. \tag{1.75}$$

The operator A being compact, the spectrum of S consists of a denumerable set of point eigenvalues, $e^{2i\delta_n}$, $n = 1, ..., \infty$, that accumulate at 1. The δ_n, which may be defined to accumulate at zero, are called the *eigenphase shifts*. We then clearly have

$$\delta = \sum_{n=1}^{\infty} \delta_n \quad (\text{mod } \pi), \tag{1.76}$$

each δ_n included as many times as the multiplicity of the corresponding eigenvalue. If we now define

$$\eta(k) := -\delta(k) - \frac{k}{4\pi}\langle V \rangle,$$

then (1.74) implies that η may be chosen to be

$$\eta(k) = \arg \Delta(k). \tag{1.77}$$

The function Δ is the boundary value of an analytic function that is holomorphic in $\mathbb{C}^+$ and it tends to 1 as $|k| \to \infty$ (Lemma 1.4.2). Furthermore, it has zeros in $\mathbb{C}^+$ at those points $i\kappa$ for which $-\kappa^2$ is a bound state, with the multiplicity of each zero equal to the multiplicity of the corresponding eigenvalue. On the real axis, Δ is continuous and without zeros, unless $k = 0$ is an exceptional point. Therefore the *argument principle*, together with the symmetry $\Delta(-k) = \overline{\Delta(k)}$, implies that

$$\eta(0) - \eta(\infty) = -\pi n,$$

where n is the number of bound states, counting each as many times as its multiplicity. If $k = 0$ is an exceptional point then n includes the dimensionality of the L^2-eigenspace with eigenvalue zero, and if there is a half-bound state (see Section 1.2) then n is replaced by $n + 1/2$. (That's the origin of the terminology.) We therefore have the following.

Lemma 1.5.9. (Generalized Levinson Theorem) *If δ is defined by (1.75) as a continuous function, $\mathbb{R}_+ \mapsto \mathbb{R}$, then*

$$\delta(0) - \lim_{k \to \infty} \left[\delta(k) + \frac{k}{4\pi} \langle V \rangle \right] = \pi(n + \tfrac{1}{2}q), \tag{1.78}$$

where $\langle V \rangle$ is defined in Lemma 1.5.8, n equals the dimension of the orthogonal complement of the absolutely continuous subspace of $H = V - \Delta$, i.e., the total number of linearly independent bound-state eigenfunctions, and $q = 1$ if there is a half-bound state at $k = 0$; $q = 0$ otherwise.

It should be noted that for $k = 0$ the Fredholm determinant $\Delta(0)$ is real because the kernel of the Lippmann-Schwinger equation is real. Therefore, if $k = 0$ is not exceptional then (1.74) implies that $\det S(0) = 1$. If $k = 0$ is exceptional of the second kind, i.e., there is half-bound state, then $\det S(0) = -1$; if it is of the first kind only then $\det S(0) = 1$. Consequently, $\delta(0) = \pi(m + 1/2q)$, where m is an integer. Comparison with (1.78) shows that therefore there must exist an integer p such that

$$\lim_{k \to \infty} \left[\delta(k) + \frac{k}{4\pi} \langle V \rangle \right] = \pi p.$$

This integer may be chosen to be zero, thereby defining $\delta(k)$ uniquely. With that choice (1.78) reads more simply ·

$$\delta(0) = \pi(n + \tfrac{1}{2}q). \tag{1.79}$$

The Generalized Levinson Theorem should be compared with the situation when V is *central*, i.e., it is a function of $|x|$ only. In that case the Schrödinger equation is separable, the eigenphase shifts δ_l are the well-known phase shifts, and the eigenvalue $e^{2i\delta_l}$ of S has the multiplicity $2l+1$. The integer l now denotes the angular momentum. There is, in that case, a separate Levinson theorem for each l, connecting the phase shift δ_l to the number of bound states of the same angular momentum. In other words, for central potentials there are infinitely many "micro-Levinson theorems," whereas for noncentral potentials Lemma 1.5.9 constitutes only one "macro-Levinson theorem," which is much less restrictive. Recently, a more powerful set of "micro-Levinson theorems" has been proved. To begin with, it was proved in [Ne89a,b] that the eigenphase shifts may be labelled by the pair of indices (l, n), $l = 1, \dots, \infty$, $n = 1, \dots, 2l + 1$, in the sense that as $k \to 0$, the eigenfunction $\mathfrak{s}_{ln}$ that corresponds to the eigenphase shift δ_{ln} approaches a (non-conventional) spherical harmonic $\mathfrak{Y}_{ln}$ of order l, which is connected to the conventional spherical harmonics Y_l^m of the same order by a unitary $(2l + 1) \times (2l + 1)$ matrix that depends on the potential. Furthermore, each bound state may also be labelled by a pair (l, n) in the sense that as the potential strength decreases to a threshold at which the bound-state eigenvalue tends to zero, the character of that bound state approaches a multiple of one of the above-mentioned spherical harmonics $\mathfrak{Y}_{ln}$. This establishes a one-to-one correspondence between the eigenphase shifts and the bound states. The following "micro-Levinson theorem" was then proved in [Ne89a,b]:

Lemma 1.5.10. *Suppose that $V \in \mathcal{W}$ and $\exists \epsilon > 0$ such that*

$$\int_{\mathbb{R}^3} dx\, |V(x)| e^{\epsilon|x|} < \infty. \tag{1.80}$$

Then each eigenphase shift $\delta_{ln}(k)$ may be defined to be a continuous function of k, to vanish at $k \to \infty$, and so that its value at the origin is

$$\delta_{ln}(0) = \pi(\mathcal{N}_{ln} + v),$$

where $\mathcal{N}_{ln}$ is the number of bound states associated with the pair (l,n), $v = \frac{1}{2}$ if $l = 0$ and there is a half-bound state, and $v = 0$ otherwise.

This lemma has been proved so far only under the strong assumption of exponential decrease of the potential. It does not, therefore, constitute one of the necessary conditions for the S matrix to be associated with a potential in $\mathcal{W}$. However, it is likely that it holds under much weaker conditions and may turn out to play a role in the inverse problem that is not yet fully understood.

If the potential is shifted as in (1.18) then the corresponding scattering amplitude is readily obtained from (1.19), (1.62), and a shift in the variable of integration as follows:

$$A_y(k, \theta, \theta') = A(k, \theta, \theta') e^{iky \cdot (\theta - \theta')}. \tag{1.81}$$

Let us now turn to the large-$|k|$ behavior of the scattering amplitude. If (1.8) is inserted in (1.62) we obtain

$$A(k, \theta, \theta') = B(\tau) + A_R(k, \theta, \theta'), \tag{1.82}$$

where

$$\tau := k(\theta' - \theta),$$

$$B(\tau) := -\frac{1}{4\pi} \int_{\mathbb{R}^3} dx\, V(x) e^{ik\tau \cdot x}, \tag{1.83}$$

$$A_R(k, \theta, \theta') := \int_{\mathbb{R}^3 \times \mathbb{R}^3} dx\, dy\, V(x) V(y) G^+(k, x, y) e^{ik(\theta' \cdot y - \theta \cdot x)}, \tag{1.84}$$

and G^+ is the complete Green's function. The function $B(\tau)$ is called the *Born Approximation* to the scattering amplitude. It is an old result of Zemach and Klein [ZK58, see also Fa56] that if V is in the Rollnik class and in L^1 then

$$\lim_{k \to \pm\infty} A_R(k, \theta, \theta') = 0$$

uniformly in θ and θ', $\theta, \theta' \in \mathbb{R}^3$, $|\theta| = |\theta'| = 1$. It follows that if τ (which physically is the *momentum transfer*) is kept fixed while $k \to \pm\infty$ then

$$\lim_{k \to \pm\infty, k(\theta' - \theta) = \tau} A(k, \theta, \theta') = B(\tau). \tag{1.85}$$

It should be noted that $\tau = 0$ for $\theta = \theta'$ and hence for the forward scattering amplitude

$$\lim_{k \to \pm\infty} A(k,\theta,\theta) = B(0) = -\langle V \rangle / 4\pi,$$

where $\langle V \rangle$ was defined below (1.74). Thus the forward scattering amplitude tends to a θ-independent constant which is generally nonzero.

It will be important for the inverse problem to be able to define the Fourier transform of $kA(k,\theta,\theta')$ in some useful sense. The fact that the forward scattering amplitude approaches a constant as $k \to \pm\infty$ shows that this sense cannot be an L^1 or L^2-meaning pointwise in θ and θ', and a distribution sense will make later results cumbersome.

For the operator-valued function $A(k)$ defined by the integral kernel $A(k,\theta,\theta')$ one can readily see that it will generally not be true that $k\|A(k)\|_2$ is in L^2. However, the following result has recently been proved in [We89].

Lemma 1.5.11. (Weder) *If $V \in \mathcal{W}$ then $\exists C \in \mathbb{R}_+$ such that for all $k \in \mathbb{R}_+$*

$$\|kA(k)\| \le \frac{C}{1+k},$$

where $\|A\|$ is the operator norm for A regarded as an operator $L^2(\mathbb{S}^2) \mapsto L^2(\mathbb{S}^2)$.

This lemma will be proved in Section 1.7.

Definition 1.5.12. $\mathcal{L}^2$ is the set of functions $f(k)$ with values that are operators $L^2(\mathbb{S}^2) \mapsto L^2(\mathbb{S}^2)$ such that $\|f(k)\| \in L^2(\mathbb{R})$, where $\| \cdot \|$ is the operator norm.

Corollary 1.5.13. *If $V \in \mathcal{W}$ then $kA(k) \in \mathcal{L}^2$ and so is its Fourier transform, defined in the $\mathcal{L}^2$-sense.*

Finally, we will later need a certain additional property that at this point may appear unmotivated and artificial. We define

$$G(\alpha,\theta,\theta') := \frac{i}{(2\pi)^2} \int_{-\infty}^{\infty} dk\, kA(k,-\theta,\theta')e^{-ik\alpha} \tag{1.86}$$

and the operator $\mathcal{G}$ whose kernel is given by

$$\mathcal{G}(\alpha,\theta;\beta,\theta') := G(\alpha+\beta;\theta,\theta'), \quad \alpha,\beta \in \mathbb{R}_+, \quad \theta,\theta' \in \mathbb{S}^2, \tag{1.87}$$

so that $(\mathcal{G}f)(\alpha,\theta) = \int_0^{\infty} d\beta \int_{\mathbb{S}^2} d\theta'\, G(\alpha+\beta,\theta,\theta')f(\beta,\theta')$. Similarly we define $\mathcal{G}^*$ by the kernel $G(-\alpha-\beta,\theta,\theta')$,

$$\mathcal{G}^*(\alpha,\theta;\beta,\theta') := G(-\alpha-\beta,\theta\theta'), \quad \alpha,\beta \in \mathbb{R}_+, \quad \theta,\theta' \in \mathbb{S}^2. \tag{1.88}$$

We then have the following result, which will be proved in Section 1.7.

Lemma 1.5.14. *If $V \in \mathcal{W}$ and A is the corresponding scattering amplitude then the operators $\mathcal{G}$ and $\mathcal{G}^*$ defined by (1.86), (1.87), and (1.88) are bounded and self-adjoint as operators $L^2(\mathbb{R}_+ \times \mathbb{S}^2) \mapsto L^2(\mathbb{R}_+ \times \mathbb{S}^2)$, and $\|\mathcal{G}^2\|_2 < \infty$ and $\|\mathcal{G}^{*2}\|_2 < \infty$. Here $\| \cdot \|_2$ is the Hilbert-Schmidt norm.*

Remark 1. If V is allowed to be complex then most of the results of this chapter will still hold. The results that fail are those that concern the point spectrum (as H is no longer self-adjoint), the completeness, and the unitarity of the S matrix.

Remark 2. If V is assumed to have compact support then one easily sees that $\psi^+(k, \theta, x)$ is an analytic function of k that is meromorphic in all of $\mathbb{C}$, and so is $A(k, \theta, \theta')$, and $\Delta(k)$ is entire. However, the zeros of $\Delta(k)$ in $\mathbb{C}^-$ have no significance for the point spectrum of H, and neither Δ nor ζ will approach 1 as $|k| \to \infty$ in $\mathbb{C}^-$. Exponential decrease of the potential results in an extension of the region of analyticity of Δ and ψ^+ to a strip below the real line, and hence analyticity of $A(k, \theta, \theta')$ in a strip that includes the real axis.

Remark 3. If $V(x)$ is replaced by a so-called *nonlocal potential*, i.e.,

$$V(x)f(x) \;\to\; \int_{\mathbb{R}^3} dy\, V(x, y) f(y),$$

then all analyticity properties of ψ^+ will generally be lost. However, $\Delta(k)$ will, for a large class of kernels $V(x, y)$, still be analytic in $\mathbb{C}^+$ and there is a version of Levinson's theorem [Ne77b, Dr76].

Remark 4. The Schrödinger equation (1.1) is obtainable by Fourier transformation with respect to k^2 from the time-dependent Schrödinger equation

$$i\frac{\partial}{\partial t}\psi = (V - \Delta)\psi,$$

or by Fourier transformation with respect to k from the so-called *plasma wave equation*:

$$\left(\Delta - \frac{\partial^2}{\partial t^2}\right) f = V f.$$

However, there appears to be a physically sensible scattering theory for the latter only if (1.1) has no bound-state eigenvalues [Ne 85c]. The *variable velocity wave equation*

$$\left(\Delta - \frac{1}{c^2}\frac{\partial^2}{\partial t^2}\right) f = 0,$$

where $c = c(x)$, léads, by Fourier transformation to

$$(k^2 V - \Delta)\psi = k^2\psi,$$

where $V = 1 - 1/c^2$, i.e., an equation like (1.1), except that V is replaced by $k^2 V$.

Remark 5. If $V(x)$ is replaced by a function that depends on k (a particularly relevant case would be $k^2 V(x)$, which arises by Fourier transformation from the wave equation with variable velocity, as in Remark 4) then all the results of this Section that do not refer to large-$|k|$ asymptotics or completeness still hold. This can be seen by simply replacing $k^2 V$ by $V_0 := k_0^2 V$, so that V_0 is independent of k and agrees with $k^2 V$ at $k = k_0$. Thus, particularly, unitarity and Lemma 1.5.1 will hold even in this case. However, the large-$|k|$ asymptotics will be quite different and is at this time not fully known.

Remark 6. In two dimensions most of the results of this Section also hold, except that the behavior of the scattering solution and of the scattering amplitude at the origin is more complicated. The reason is that the Green's function has a logarithmic singularity at $k = 0$. Most of the relevant results known can be found in references [Ch82, 84a, 84b, and 85]. In dimensions higher than three the results for large $|k|$ along the real axis break down and what analogous results take their place is unknown. (See, however, [We89].)

Remark 7. The "impedance Schrödinger equation"

$$\alpha^{-2}\nabla \cdot [\alpha^2\nabla\psi] + k^2\psi = V\psi, \quad x \in \mathbb{R}^3,$$

in which $\alpha \in \mathbb{R}$ and $\nabla\alpha$ are allowed to have discontinuities on a finite number of nested closed surfaces, is of considerable practical interest in acoustics. Much progress has recently been made by P. C. Sabatier in developing a scattering theory for this equation [see Sb87b, 88, 89a,b,c,d].

Remark 8. The scattering problem of the Schrödinger equation for an electrically charged particle in a magnetic field,

$$[\nabla + i\mathfrak{A}(x)]^2 \psi + k^2\psi = V(x)\psi,$$

$\mathfrak{A} \in \mathbb{R}^3$, $V \in \mathbb{R}$, has been considered by R. G. Novikov and G. M. Henkin in [NH87b, see also HN88].

In summing up the results of this Section we will use the operator language for the scattering amplitude and the S matrix, and also introduce the operator Q that acts on functions on $\mathbf{S}^2$:

$$(Qf)(\theta) := f(-\theta).$$

We define a set of classes of functions, $\mathbb{R} \times \mathbf{S}^2 \times \mathbf{S}^2 \mapsto \mathbb{C}$ that are regarded as integral kernels of operators with the following properties.

Definition 1.5.15. $A \in \mathscr{A}$ if

(i) the kernel $A(k, \theta, \theta')$ that defines the operator family $A(k)$ is a continuous, uniformly bounded, differentiable function $\mathbb{R} \times \mathbf{S}^2 \times \mathbf{S}^2 \mapsto \mathbb{C}$,

(ii) $QAQ = \tilde{A}$ *(reciprocity)*, (the tilde here means the operator whose kernel is the transpose)

(iii) $A(-k) = \overline{A(k)}$,

(iv) $S := \mathbb{1} - \frac{k}{2\pi i}A$, $S^\dagger S = SS^\dagger = \mathbb{1}$, *(unitarity)*,

(v) $kA(k) \in \mathscr{L}^2$, where $\mathscr{L}^2$ is defined in Definition 1.5.12.

(vi) the operators $\mathscr{G}$ and $\mathscr{G}^*$ defined by (1.86), (1.87), and (1.88) in terms of A are such that $\mathscr{G}^2$ and $\mathscr{G}^{*2}$ are Hilbert-Schmidt.

The subclasses $\mathscr{A}_n$ are defined by adding the following two requirements:

(vii) $\delta := \frac{1}{2}\arg\det S$, defined as a continuous function of k, is such that $\exists\delta_\infty$ and

$$\delta(0) - \lim_{k\to\infty}[\delta(k) + k\delta_\infty] = \pi n,$$

(Generalized Levinson Theorem)

(viii) $A(k, \theta, \theta)$ is, for each $\theta \in \mathbf{S}^2$, the boundary value of an analytic function meromorphic in $\mathbb{C}^+$ with simple poles at points $k = i\kappa_m$ on the positive imaginary axis *(forward analyticity)* which are all *normal* in the sense of Definition 1.5.4. The integers N'_m defined there for poles at $i\kappa_m$ are such that $n = \sum_m N'_m$.

This definition of $\mathscr{A}_n$ explicitly rules out that $k = 0$ is an exceptional point, and it requires that if there are bound states, they all be normal. The results of this Section can be summarized in the following.

Theorem 1.5.16. *If* $V \in \mathcal{W}$ *and* $H = V - \Delta$ *has no bound or half-bound states then* $A \in \mathcal{A}_0$. *If*

(i) $V \in \mathcal{W}$ *and* $(1 + |x|^{N_0})V \in L^1(\mathbb{R}^3)$,

(ii) *H has b bound states of eigenvalues* $-\kappa_m^2$ *with degeneracy* N_m, $m = 1, \ldots, b$,

(iii) *these are all normal in the sense of Definition 1.2.5,*

(iv) $k = 0$ *is not an exceptional point,*

then $A \in \mathcal{A}_n$, *where* $n = \sum_{m=1}^b N_m$ *and* $N_m = N_m'$ *for each m. The number* N_0 *in the hypothesis equals the largest of the numbers* N_m.

Note that if $V \in \mathcal{W}$ then the translated potential $V_y \in \mathcal{W}$ for all $y \in \mathbb{R}^3$; recall also that (see Definition 1.1.2) $\mathcal{V}_0 \subset \mathcal{W}$.

1.6 Angular Momentum Projections

For central potentials the Schrödinger equation is separable and, physically, angular momentum is conserved. In the noncentral case, of course, it is not. Nevertheless, for some purposes it is useful to expand the dependence of the scattering solution on the direction θ of the incident wave on the basis of spherical harmonics. In order to simplify the notation as much as possible we shall use the single subscript L to denote the pair of integers usually called l, m, $l = 1, 2, \ldots$, $-l \leq m \leq l$, and we denote the usual normalized spherical harmonics by $Y_L(\theta)$ [see, e.g., Ne82c, p.31].

Define

$$\psi_L^+(k, x) := \int_{S^2} d\theta \, \overline{Y_L(\theta)} \psi^+(k, \theta, x),$$

which certainly exists for each L if $V \in \mathcal{W}$. It then follows from the Lippmann-Schwinger equation (1.3) and the expansion of the plane wave

$$e^{ik\theta \cdot x} = 4\pi \sum_L i^l j_l(kr) Y_L(\theta) \overline{Y_L(\hat{x})},$$

that ψ_L^+ satisfies the integral equation

$$\psi_L^+(k, x) = 4\pi i^l j_l(kr) \overline{Y_L(\hat{x})} - \frac{1}{4\pi} \int_{\mathbb{R}^3} dy \, \frac{e^{ik|x-y|}}{|x - y|} V(y) \psi_L^+(k, y), \tag{1.89}$$

where $r = |x|$, $\hat{x} = x/|x|$, and j_l is the spherical Bessel function. This equation is solvable by Fredholm methods if $V \in \mathcal{W}$. Furthermore, if V satisfies the additional requirement

$$\int_{\mathbb{R}^3} dx \, |V(x)| \, |x|^l < \infty, \tag{1.90}$$

and the Fredholm determinant does not vanish for $k = 0$, then one easily finds that as $k \to 0$

$$\psi_L^+(k, x) = O(k^l). \tag{1.91}$$

Therefore, if there exists some $\epsilon > 0$ such that (1.80) holds, then (1.90) is certainly satisfied for all l and hence (1.91) holds for all l. Furthermore, if (1.80) holds and

$k = 0$ is not an exceptional point, then both $\psi^+(k, \theta, x)$ and $\psi_L^+(k, x)$ are analytic as functions of k at $k = 0$. In addition to (1.91) one readily shows by means of the integral equation (1.89) that for each L, $\exists C$ such that

$$\| \, |V(\cdot)|^{1/2} \psi_L^+(k, \cdot) \| < C|k|^l, \tag{1.92}$$

where $\| \cdot \|$ is the norm on $L^2(\mathbb{R}^3)$.

The next step is to define similar expansion coefficients for the scattering amplitude in its dependence on θ and θ'

$$A_{LL'}(k) := \int_{S^2 \times S^2} d\theta d\theta' \, Y_L(\theta) \overline{Y_{L'}(\theta')} A(k, \theta, \theta'). \tag{1.93}$$

It then follows from the integral representation (1.62) that $A_{LL'}$ has the representation

$$A_{LL'}(k) = -(-i)^l \int_{\mathbb{R}^3} dx \, V(x) \overline{Y_L(\theta)} j_l(kr) \psi_{L'}^+(k, x). \tag{1.94}$$

One easily proves by means of (1.92) and the inequality

$$|j_l(z)| \leq C_l \frac{|z|^l}{(1 + |z|)^{l+1}}$$

the following result.

Lemma 1.6.1. *If the potential is in $\mathcal{W}$ and satisfies (1.80) for some $\epsilon > 0$, and if $k = 0$ is not an exceptional point, then for all L and L', $A_{LL'}(k)$ is an analytic function of k that is holomorphic on the real axis and $\exists C_{ll'}$ such that for all $k \in \mathbb{R}$*

$$|A_{LL'}(k)| < C_{ll'}|k|^{l+l'}.$$

It should be noted that in the special case of central potentials, of course, $A_{LL'} = 0$ unless $L = L'$ and Lemma 1.6.1 is well known [see, e.g., Ne82c].

Corollary 1.6.2. *If the assumptions of Lemma 1.6.1 are satisfied then for all $n < l+l'$*

$$\left. \frac{\partial^n}{\partial k^n} A_{LL'}(k) \right|_{k=0} = 0.$$

There is also an immediate consequence for the Fourier transform of kA, the function G defined by (1.86). If we define

$$\begin{aligned} G_{LL'}(\alpha) &:= \int_{S^2 \times S^2} d\theta d\theta' \, Y_L(\theta) \overline{Y_{L'}(\theta')} G(\alpha, \theta, \theta') \\ &= \frac{(-1)^l}{i(2\pi)^2} \int_{-\infty}^{\infty} dk \, k e^{ik\alpha} A_{LL'}(-k) \end{aligned} \tag{1.95}$$

and

$$\rho_{LL'}^n := \frac{1}{n!} \int_{-\infty}^{\infty} d\alpha \, \alpha^n G_{LL'}(\alpha), \tag{1.96}$$

then we have the following.

Corollary 1.6.3. *If the assumptions of Lemma 1.6.1 are satisfied then for all $n \leq l + l'$, $\rho_{LL'}^n = 0$.*

1.7 Proofs

Proof of Lemma 1.2.1. We have by the differentiation rule for the modified Fredholm determinant

$$\frac{d}{dk} \log \det_2(\mathbb{1} - G_0 V) = -\text{tr}\left[(\mathbb{1} - G_0 V)^{-1} \frac{dG_0}{dk} V G_0 V\right]$$

$$= 2k \, \text{tr} \, G(G_0 V)^2,$$

where G is the full Green's function and the differentiations are justified by the absolute convergence of the series and integrals. Now let $\Delta = \det_2(\mathbb{1} - G_0 V)$ have a zero of order p at $k = k_0$, and let P be the projection onto the n-dimensional eigenspace of H at k_0^2, so that $G_0 V P = P$ there. Then

$$p = \lim_{k \to k_0} \frac{d}{dk} \log \Delta(k) = \text{tr} P (G_0 V)^2 = \text{tr} P = n$$

because G, as the resolvent of the self-adjoint operator H, has a simple pole at $k = k_0$ whose residue is P.   □

Proof of Lemma 1.2.2. For the proof of (1.21) and (1.24) see [Ne77a]. To prove (1.22) we use the fact that

$$(\mathbb{1} - L)^{-1} = \mathbb{1} + |V|^{1/2} G V^{1/2},$$

where $G = (\mathbb{1} - G_0 V)^{-1} G_0$ has the pole-term

$$G(k) = \frac{1}{k^2} \mathscr{P}_b + O(1);$$

(1.22) then follows. For (1.23) we use (1.14) and (1.22). Equation (1.25) is proved by defining $Z^2 := -G_0(0)$ and

$$B(k) := -Z^{-1} G(k) V Z = B_0 + ik Z^{-1} G_1 V Z + \ldots,$$

with the pole term

$$[\mathbb{1} - B(k)]^{-1} = \frac{1}{k} \mathscr{P}_0 + O(1), \quad \text{where}$$

$$\mathscr{P}_0(x, y) := 4\pi i \frac{\chi(x)\chi(y)}{(\int dz \, V u)^2},$$

$(\mathbb{1} - B_0)\chi = 0$, and $u = G_0(0) V u = Z \chi$. One then finds (1.25) by straight-forward computation. Similarly one obtains (1.26) using (1.14) and (1.25).   □

Proof of Lemma 1.4.2. We define the integral kernel $L(k, x, y)$ by (1.13) and the operator-valued function $L(k)$ by the kernel $L(k, x, y)$ on $\mathbb{R}^3$. It is known that [Si71, p.23] if V is in the Rollnik class then

$$\lim_{|k| \to \infty} \|L^\dagger(k) L(k)\|_2 = 0, \quad k \in \mathbb{C}^+ \cup \mathbb{R}, \tag{1.97}$$

which implies that

$$\lim_{|k| \to \infty} \|L(k)\| = 0, \tag{1.98}$$

where $\| \cdot \|$ is the operator norm on $L^2(\mathbb{R}^3)$. The operator K_0 defined in (1.6) is related to L by $K_0(k, \theta, x, y) = L(k, x, y)e^{ik\theta \cdot (y-x)}$. Therefore (1.97) implies that

$$\lim_{k \to \pm \infty} \|K_0^\dagger(k, \theta)K_0(k, \theta)\|_2 = 0 \tag{1.99}$$

uniformly for all $\theta \in S^2$. The same is easily seen to be true as $\Re k \to \pm \infty$ for any fixed $\Im k \geq 0$, as well as for $\Im k \to \infty$ by Lebesgue's dominated convergence theorem. Therefore for each $\theta \in S^2$ and $\Im k \geq 0$

$$\lim_{|k| \to \infty} \|K_0(k, \theta)\| = 0. \tag{1.100}$$

We define $h(k, \theta, x) := |V(x)|^{1/2}[\zeta(k, \theta, x) - 1]$ and obtain the integral equation from (1.3)

$$\begin{aligned} h(k, \theta, x) &= \int_{\mathbb{R}^3} dy\, K_0(k, \theta, x, y)|V(y)|^{1/2} \\ &+ \int_{\mathbb{R}^3} dy\, K_0(k, \theta, x, y)h(k, \theta, y). \end{aligned} \tag{1.101}$$

Together with (1.100) this implies that for $k \in \mathbb{C}^+ \cup \mathbb{R}$

$$\lim_{|k| \to \infty} \|h(k, \theta, \cdot)\| = 0, \tag{1.102}$$

where $\| \cdot \|$ is the norm on $L^2(\mathbb{R}^3)$. The function ζ can be obtained from h by the equation

$$\zeta(k, \theta, x) - 1 = \alpha(k, \theta, x) + \int_{\mathbb{R}^3} dy\, G_0'(k, \theta, x - y)V^{1/2}(y)h(k, \theta, y), \tag{1.103}$$

where $\alpha(k, \theta, x) := \int_{\mathbb{R}^3} dy\, G_0'(k, \theta, x - y)V(y)$ and G_0' is defined below (1.8). One obtains the desired result (1.54) from (1.103), because (1.102) and the definition of $\mathcal{V}$ imply that the second term in (1.103) tends to zero as $|k| \to \infty$. That α tends to zero follows from the definition of $\mathcal{V}$ and the dominated convergence theorem.

We now turn to the proof of (1.56), which will require (ii) and (iii) in the definition of $\mathcal{W}$.

We first estimate the function

$$\alpha(k, \theta, x) = -\int_{\mathbb{R}^3} dy \frac{V(y)}{4\pi|x - y|} e^{ik(|x-y|+\theta \cdot (y-x))}.$$

It immediately follows from assumption (i) that $|\alpha|$ is bounded uniformly for all $k \in \mathbb{R}$, $\theta, x \in \mathbb{R}^3$, $\theta \cdot \theta = 1$. Furthermore, by an integration by parts, writing $r = |y|$ after a shift in the variable of integration,

$$\begin{aligned} \alpha &= \int_0^\infty dr\, r e^{ikr} \int_{S^2} d\hat{y}\, V(x + y)e^{ik\theta \cdot y} \\ &= \frac{2\pi i}{k}\left[\int_0^\infty dr\, V(x - r\theta) - \int_0^\infty dr\, V(x + r\theta)e^{2ikr}\right] \\ &\quad + \frac{i}{k}\int_0^\infty dr\, r \int_{S^2} d\hat{y}\, \theta \cdot \nabla V(x + y)e^{ik(r + \theta \cdot y)}, \end{aligned}$$

and therefore by (ii) and (iii), with $s > 1$,

$$|\alpha| \le \frac{C}{|k|} \left[\int_0^\infty dr \, [1 + (r - |x|)^2]^{-s} + \int_{\mathbb{R}^3} dy \frac{|\nabla V(y)|}{|x - y|} \right] \le \frac{C'}{|k|}.$$

Consequently $\exists C$ such that for all $k \in \mathbb{R}$, $\theta, x \in \mathbb{R}^3$, $\theta \cdot \theta = 1$

$$|\alpha(k, \theta, x)| \le \frac{C}{1 + |k|}. \tag{1.104}$$

From (1.103) we have

$$|\zeta(k, \theta, x) - 1| \le |\alpha(k, \theta, x)| + \int_{\mathbb{R}^3} dy \frac{|V(y)|^{1/2}}{4\pi|x - y|} |h(k, \theta, y)|$$

and by Schwarz's inequality,

$$\int dk \left(\int dy \frac{|V|^{1/2}}{|x - y|} |h| \right)^2$$

$$= \int dy \, dz \frac{|V(y) V(z)|^{1/2}}{|x - y||x - z|} \int dk \, |h(k, \theta, y) h(k, \theta, z)|$$

$$\le \left\{ \int dy \frac{|V(y)|^{1/2}}{|x - y|} \left[\int dk \, |h(k, \theta, y)|^2 \right]^{1/2} \right\}^2$$

$$\le \int dy \frac{|V(y)|}{|x - y|^2} \int dz \int dk \, |h(k, \theta, z)|^2$$

$$\le C \int_{-\infty}^\infty dk \, \|h(k, \theta, \cdot)\|^2,$$

because of (i). We therefore have for all $\theta \in \mathbf{S}^2$ and $x \in \mathbb{R}^3$

$$\int_{-\infty}^\infty dk \, |\zeta(k, \theta; x) - 1|^2 \le C + C' \int_{-\infty}^\infty dk \, \|h(k, \theta, \cdot)\|^2.$$

We must now estimate $\|h(k, \theta, \cdot)\|$.
From (1.101)

$$h(k, \theta, x) = \int_{\mathbb{R}^3} dy \, (\mathbb{1} - K_0)^{-1}(x, y) |V(y)|^{1/2} \alpha(k, \theta, y),$$

and therefore

$$\|h(k, \theta, \cdot)\| \le \|(\mathbb{1} - K_0)^{-1}\| \, \|V^{1/2} \alpha(k, \theta, \cdot)\|.$$

If $k = 0$ is not an exceptional point then $\|(\mathbb{1} - K_0)^{-1}\|$ is bounded for each real k and by (1.100) it tends to 1 as $k \to \pm\infty$. Therefore, by (i),

$$\int_{-\infty}^\infty dk \, \|h(k, \theta, \cdot)\|^2 \le C \int_{-\infty}^\infty dk \, \|V^{1/2}(\cdot)\alpha(k, \theta, \cdot)\|^2$$

$$= C \int_{\mathbb{R}^3} dx \, |V(x)| \int_{-\infty}^\infty dk \, |\alpha(k, \theta, x)|^2 < \infty.$$

This proves (1.56) if $k = 0$ is not an exceptional point. If it is, it is merely necessary to remove an interval around the origin from the k-integral.

Next we must deal with the Fredholm determinant $\det_2(\mathbb{1} - L)$, which may be expressed as a power series in which each term is a polynomial in $\mathrm{tr}L^n$. We have

$$|\mathrm{tr}L^n| \leq \|L^2\|_2 \|L^{n-2}\|_2 \leq \|L^2\|_2^2 \|L^{n-4}\| \leq \|L^2\|_2^2 \|L\|^{n-4},$$

where $\| \cdot \|$ is the operator norm and $\| \cdot \|_2$ is the Hilbert-Schmidt norm. Furthermore,

$$\|L^2\|_2^2 = \mathrm{tr}LL^\dagger L^\dagger L \leq [\mathrm{tr}(LL^\dagger)^2]^{1/2}[\mathrm{tr}(L^\dagger L)^2]^{1/2} = \mathrm{tr}(LL^\dagger)^2 = \|L^\dagger L\|_2,$$

and therefore for $n \geq 4$

$$|\mathrm{tr}L^n| \leq \mathrm{tr}(LL^\dagger)^2 \|L\|^{n-4}. \tag{1.105}$$

For $n = 3$ we have

$$\mathrm{tr}L^3 = \frac{-1}{(4\pi)^3} \int_{\mathbb{R}^3 \times \mathbb{R}^3 \times \mathbb{R}^3} dx\, dy\, dz\, f(x, y, z) e^{ik(|x-y|+|y-z|+|z-x|)},$$

where

$$f(x, y, z) := \frac{V(x)V(y)V(z)}{|x-y||y-z||z-x|}.$$

Now by Schwarz's inequality

$$\int dx\, dy\, dz\, |f(x, y, z)|$$

$$\leq \int dx\, dy \frac{|V(x)||V(y)|}{|x-y|} \left[\int dz \frac{|V(z)|}{|x-z|^2}\right]^{1/2} \left[\int dw \frac{|V(w)|}{|y-w|^2}\right]^{1/2}$$

$$\leq C \int dx\, |V(x)| \left[\int dy\, dz \frac{|V(y)||V(z)|}{|y-z|^2}\right]^{1/2} < C'$$

by the assumption (*i*). Therefore $f \in L^1(\mathbb{R}^9)$, and it follows, as in the argument of [Si71, p. 24], that

$$\lim_{|k| \to \infty} \mathrm{tr}L^3 = 0$$

for $\Im k \geq 0$, either by Riemann-Lebesgue or by dominated convergence.

Finally we have

$$\mathrm{tr}L^2 = \int_{\mathbb{R}^3 \times \mathbb{R}^3} dx\, dy \frac{V(x)V(y)}{|x-y|^2} e^{2ik|x-y|},$$

which is easily seen to vanish as $|k| \to \infty$ if V is in the Rollnik class. The result (1.55) therefore follows from the expansion

$$\Delta(k) = \exp\left(-\sum_{n=2}^\infty \frac{1}{n} \mathrm{tr}L^n(k)\right)$$

which by (1.98) and (1.105) converges for $|k|$ sufficiently large.   □

Proof of Lemma 1.5.8. We start by noting that for the modified Fredholm determinant

$$\det_2[(\mathbb{1} - A)(\mathbb{1} - B)] = \det_2(\mathbb{1} - A)\det_2(\mathbb{1} - B)e^{\mathrm{tr}(\mathbb{1}-A)B}. \tag{1.106}$$

Therefore,

$$
\begin{aligned}
\overline{\Delta(k)} &= \det_2(\mathbb{1} - G_0^- V) \\
&= \det_2\{(\mathbb{1} - G_0^+ V)[\mathbb{1} - (\mathbb{1} - G_0^+ V)^{-1}(G_0^- - G_0^+)V]\} \\
&= \det_2(\mathbb{1} - G_0^+ V)\exp\{2\pi i\,\mathrm{tr}[V\delta(k^2 - H_0)]\} \\
&\quad \times \det\{\mathbb{1} - (\mathbb{1} - G_0^+ V)^{-1}2\pi i\delta(k^2 - H_0)V\},
\end{aligned}
$$

where $\delta(k^2 - H_0)$ is the analogue of the kernel defined by (1.49). Thus by the Lippmann-Schwinger equation the operator $(\mathbb{1} - G_0^+ V)^{-1}\delta(k^2 - H_0)V$ has the kernel $-kM(x,y)/(2\pi)^2$, where

$$M(x,y) = -\frac{1}{4\pi}\int_{S^2} d\theta\,\psi^+(k,\theta,x)V(y)e^{-ik\theta\cdot y}.$$

It is easily seen that $\mathrm{tr}M^n = \mathrm{tr}A^n$, where A is the scattering amplitude and the trace on the right is over the unit sphere. Therefore,

$$\det\{\mathbb{1} - (\mathbb{1} - G_0^+ V)^{-1}2\pi i\delta(k^2 - H_0)V\} = \det\left(\mathbb{1} - \frac{k}{2\pi i}A\right) = \det S.$$

Furthermore,

$$\mathrm{tr}V\delta(k^2 - H_0) = \frac{k}{(2\pi)^2}\langle V\rangle,$$

and the result is (1.74).   □

Proof of Lemma 1.5.11. We will work in the weighted L^2-space defined by

$$L_s^2(\mathbb{R}^3) = \left\{f : \mathbb{R}^3 \mapsto \mathbb{C} \,\middle|\, \|f\|_s^2 := \int_{\mathbb{R}^3} dx\,|f(x)|^2(1 + |x|^2)^s < \infty\right\}.$$

Also define the function $\sigma(k)$, $k \in \mathbb{R}_+$, with values that are operators $L_s^2(\mathbb{R}^3) \mapsto L^2(S^2)$ with $s > 1$, by

$$h(k\theta) := (\sigma(k)f)(\theta) := \int_{\mathbb{R}^3} dx\,f(x)e^{ik\theta\cdot x}.$$

The adjoint $\sigma^\dagger(k)$ of $\sigma(k)$ is an operator $L^2(S^2) \mapsto L_{-s}^2(\mathbb{R}^3)$ given by

$$\phi(kx) := (\sigma^\dagger(k)\varphi)(x) = \int_{S^2} d\theta\,\varphi(\theta)e^{-ik\theta\cdot x}.$$

The main tool is the following lemma.

Lemma 1.7.1. $\exists C$ *such that for all* $k \in \mathbb{R}_+$

$$\|\sigma(k)\|_a^2 \leq \frac{C}{k(1+k)}, \quad \|\sigma^\dagger(k)\|_b^2 \leq \frac{C}{k(1+k)},$$

where $\|\cdot\|_a$ *is the operator norm,* $L_s^2(\mathbb{R}^3) \mapsto L^2(S^2)$, *and* $\|\cdot\|_b$ *is the operator norm,* $L^2(S^2) \mapsto L_{-s}^2(\mathbb{R}^3)$, *for* $s > 1$.

Proof. Define $g(k) := \int_{S^2} d\theta \, |h(k\theta)|^2 \geq 0$. Then

$$k_0^2 g(k_0) = \int_0^{k_0} dk(k^2 g' + 2kg) \leq \int_0^\infty dk(k^2|g'| + 2kg),$$

where the prime on g denotes the derivative. Now we have $\partial h/\partial k = \theta \cdot \nabla h$, $|\partial h/\partial k|^2 \leq |\nabla h|^2$, $\nabla h = i \int dx \, x f e^{ik\theta \cdot x}$, and

$$|g'(k)| = 2 \left| \Re \int_{S^2} d\theta \, \bar{h} \frac{\partial h}{\partial k} \right| \leq 2 \left[\int_{S^2} d\theta \, |h|^2 \right]^{1/2} \left[\int_{S^2} d\theta \left| \frac{\partial h}{\partial k} \right|^2 \right]^{1/2},$$

so that

$$\int_0^\infty dk \, k^2 |g'| \leq 2 \int_0^\infty dk \, k^2 \left[\int_{S^2} d\theta \, |h|^2 \right]^{1/2} \left[\int_{S^2} d\theta' \left| \frac{\partial h}{\partial k} \right|^2 \right]^{1/2}$$

$$\leq 2 \left[\int_0^\infty dk \, k^2 \int_{S^2} d\theta \, |h|^2 \right]^{1/2} \left[\int_0^\infty dk \, k^2 \int_{S^2} d\theta' \left| \frac{\partial h}{\partial k} \right|^2 \right]^{1/2}.$$

But

$$\int_0^\infty dk \, k^2 \int_{S^2} d\theta \left| \frac{\partial h}{\partial k} \right|^2 \leq \int_0^\infty dk \, k^2 \int_{S^2} d\theta \, |\nabla h|^2$$

$$= (2\pi)^3 \int_{\mathbb{R}^3} dx \, |x|^2 |f|^2 \leq C\|f\|_1^2.$$

Hence

$$\int_0^\infty dk \, k^2 |g'| \leq C\|f\|_0 \|f\|_1 \leq C\|f\|_s^2.$$

The other integral is split into two parts: $\int_0^\infty dk \, kg = \int_0^1 + \int_1^\infty$. In the first part we use

$$kg = 4\pi \int_{\mathbb{R}^3} dx \, f(x) \int_{\mathbb{R}^3} dy \, \overline{f(y)} \frac{\sin k|x-y|}{|x-y|} \leq 4\pi \|f\|_s^2 I^{1/2},$$

where

$$I = \int_{\mathbb{R}^3 \times \mathbb{R}^3} dx \, dy \, \frac{(1+|x|^2)^{-s}(1+|y|^2)^{-s}}{|x-y|^2} < \infty$$

if $s > 1$. Therefore,

$$\int_0^1 dk \, kg \leq C\|f\|_s^2,$$

and

$$\int_1^\infty dk\,kg \le \int_1^\infty dk\,k^2 g = (2\pi)^3 \|f\|_0^2 \le C\|f\|_s^2,$$

and it follows that $k^2 g(k) \le C\|f\|_s^2$ for $s > 1$. Since we furthermore found that $kg \le C\|f\|_s^2$, we may conclude that

$$g(k) \le \frac{C\|f\|_s^2}{k(1+k)},$$

which proves the first part of Lemma 1.7.1.

The second part follows from the first:

$$\left| \int_{\mathbb{R}^3} dx\,\overline{f(x)}\phi(kx) \right| = \left| \int_{\mathbb{S}^2} d\theta\,\varphi(\theta)\overline{h(k\theta)} \right| \le \frac{C\|\varphi\|\,\|f\|_s}{k^{1/2}(1+k)^{1/2}},$$

which implies that for all $f_1(x)$ with $\|f_1\|_0 = 1$,

$$\int_{\mathbb{R}^3} dx\,\overline{f_1(x)}\phi(kx)(1+|x|^2)^{-s/2} \le \frac{C\|\varphi\|}{k^{1/2}(1+k)^{1/2}}.$$

Since

$$\|f_2\|_0 = \sup_{\|f\|_0=1} \left| \int_{\mathbb{R}^3} dx\,f_2(x)\overline{f(x)} \right|,$$

it follows that

$$\|\phi(k\cdot)\|_{-s} \le \frac{C\|\varphi\|}{k^{1/2}(1+k)^{1/2}},$$

which implies the second part of Lemma 1.7.1.   □

We now return to the proof of Lemma 1.5.11. In view of (1.62) the scattering amplitude may be written in the operator form

$$A(k) = -\frac{1}{4\pi}\sigma(k)V^{1/2}[\mathbb{1} - L(k)]^{-1}|V|^{1/2}\sigma^\dagger(k),$$

where $L(k)$ is the operator, $L^2(\mathbb{R}^3) \mapsto L^2(\mathbb{R}^3)$, defined in (1.13). The operator $[\mathbb{1} - L(k)]^{-1}$ is uniformly bounded for all $k \ge k_0 > 0$. Furthermore, it follows from item (ii) in the definition of $\mathscr{W}$ that $V^{1/2}$ and $|V|^{1/2}$ may be regarded as bounded operators $L_{-s}^2(\mathbb{R}^3) \mapsto L^2(\mathbb{R}^3)$ and they may also be regarded as bounded operators $L^2(\mathbb{R}^3) \mapsto L_s^2(\mathbb{R}^3)$, for $s > 1$. Therefore Lemma 1.7.1 implies that $k\|A\| \le C/(1+k)$ for all $k \ge k_0 > 0$, where $\|\cdot\|$ is the operator norm for A regarded as an operator $L^2(\mathbb{S}^2) \mapsto L^2(\mathbb{S}^2)$. For $0 \le k \le k_0$, on the other hand, the same follows from the unitarity of $S = \mathbb{1} - \frac{k}{2\pi i}A$. This proves the lemma.   □

Proof of Lemma 1.5.14. We split the scattering amplitude into several terms to be considered separately:

$$\begin{aligned} A &= A^{(0)} + A', \\ A' &= A^{(1)} + A^{(2)} + A^{(R)}, \end{aligned}$$

and correspondingly,

$$
\begin{aligned}
G &= G^{(0)} + G', \\
G' &= G(1) + G^{(2)} + G^{(R)}, \\
\mathcal{G} &= \mathcal{G}^{(0)} + \mathcal{G}', \\
\mathcal{G}' &= \mathcal{G}^{(1)} + \mathcal{G}^{(2)} + \mathcal{G}^{(R)},
\end{aligned}
$$

where

$$
A^{(0)}(k, \theta, \theta') = -\frac{1}{4\pi} \int_{\mathbb{R}^3} dx\, V(x) e^{ikx \cdot (\theta' - \theta)},
$$

$$
A^{(n)}(k, \theta, \theta') = -\frac{1}{4\pi} \int_{\mathbb{R}^3 \times \mathbb{R}^3} dx\, dy\, V(x) [G_0^+(k)V]^n (x, y) e^{ik(y \cdot \theta' - x \cdot \theta)},
$$

$$
n = 1, 2,
$$

$$
A^{(R)}(k, \theta, \theta') = -\frac{1}{4\pi} \int_{\mathbb{R}^3 \times \mathbb{R}^3} dx\, dy
$$
$$
\times V^{1/2}(x) |V|^{1/2}(y) \{ [\mathbb{1} - L(k)]^{-1} L^3(k) \}(x, y) e^{ik(y \cdot \theta' - x \cdot \theta)},
$$

in which L is defined by (1.13). The corresponding terms in G and $\mathcal{G}$ are defined by Fourier transformation as in (1.86) and (1.87).

Lemma 1.7.2. *Suppose that $V \in \mathcal{W}$. (Actually, only (i) and (ii) are needed.) Then for each $x, y \in \mathbb{R}^3$ the function*

$$
I(k, x, y) := \int_{\mathbb{R}^3} dz\, G_0^+(k, x - z) V(z) G_0^+(k, z - y)
$$

is uniformly bounded as a function of k, $-\infty < k < \infty$,

$$
\lim_{|k| \to \infty} I(k, x, y) = 0, \ k \in \mathbb{C}^+ \cup \mathbb{R}, \tag{1.107}
$$

uniformly in x and y, and $\exists C$ such that for all $x, y \in \mathbb{R}^3$

$$
\int_{-\infty}^{\infty} dk\, |I(k, x, y)|^2 \le C.
$$

Proof.

$$
(4\pi)^2 I(k, x, y) = \int_{\mathbb{R}^3} dz\, V(z) \frac{e^{ik(|x-z|+|y-z|)}}{|x - z||y - z|}.
$$

It follows from Schwarz's inequality that if $V \in \mathcal{V}$ then the integrand of I is in $L^1(\mathbb{R}^3)$ and $I(k, x, y)$ is uniformly bounded for all $k \in \mathbb{R}$ and $x, y \in \mathbb{R}^3$, and by the same argument as in [Si71, p.23], (1.107) is obtained by the Riemann–Lebesgue lemma. For $\Im k \to \infty$ the same follows by dominated convergence. The square–integrability is proved as follows.

Shifting variables of integration and setting $w = y - z$, we get

$$
(4\pi)^2 I(k, x, y) = \int_{\mathbb{R}^3} dz\, V(z + x) \frac{e^{ikt}}{|z||z - w|},
$$

where $t = |z| + |z - w|$. Thus

$$(4\pi)^2 I = \frac{1}{2} \int_{|w|}^{\infty} dt e^{ikt} \int_{\mathbf{S}^2} d\hat{z} \, \frac{t^2 - w^2}{(t - w \cdot \hat{z})^2} V(x + \hat{z}r),$$

where $\hat{z} = z/|z|$ and $r = |z| = (t^2 - w^2)/2(t - w \cdot \hat{z})$. Now $r \geq (t^2 - w^2)/2(t + |w|) = (t - |w|)/2$, $|x + \hat{z}r| \geq |r - |x|| \geq |t - |w| - 2|x||/2 \geq |t - 3|x| - 3|y||/2$, and $\int_{\mathbf{S}^2} d\hat{z} \, (t^2 - w^2)/(t - w \cdot \hat{z})^2 = 4\pi$; therefore

$$\int_{-\infty}^{\infty} dk \, |I(k, x, y)|^2 \leq C \int_{|w|}^{\infty} dt \, [1 + (t - 3|x| - 3|y|)^2]^{-s} \leq C < \infty$$

with $s > 1$ by hypothesis (ii).   □

The kernel of L^2 is $L^2(k, x, y) = |V(x)|^{1/2} I(k, x, y) V^{1/2}(y)$ and we have the

Corollary 1.7.3. *For almost all* $x, y \in \mathbb{R}^3$ $L^2(k, x, y)$ *is bounded uniformly in* $-\infty < k < \infty$ *and*

$$\lim_{|k| \to \infty} L^2(k, x, y) = 0, \quad k \in \mathbb{C}^+ \cup \mathbb{R},$$

$$\int_{-\infty}^{\infty} dk \, |L^2(k, x, y)|^2 < \infty,$$

$$\int_{-\infty}^{\infty} dk \, \|L^2(k)\|_2^2 < \infty.$$

Here $\| \cdot \|_2$ *denotes the Hilbert–Schmidt norm.*

Lemma 1.7.4. $\int_0^{\infty} d\alpha \, \alpha^n \|G^{(R)}(\alpha)\|_{\mathbf{S}^2}^2 < \infty, \ n = 0, 1.$

Lemma 1.7.5. $\int_0^{\infty} d\alpha \, \alpha^n \|G^{(m)}(\alpha)\|_{\mathbf{S}^2}^2 < \infty, \ n = 0, 1, \ m = 1, 2.$

Lemma 1.7.6. $\mathcal{G}^{(0)}$ *is a bounded operator and* $\|\mathcal{G}^{(0)2}\|_2 < \infty.$

Lemma 1.7.7. *For each* $0 < \alpha < 1$ *there exists a* C *such that for all* $\theta, \theta' \in \mathbb{R}^3$, $|\theta| = |\theta'| = 1$,

$$I(\theta, \theta') := \int_{\mathbf{S}^2} d\theta'' \, |\theta - \theta''|^{\alpha-2} |\theta' - \theta''|^{\alpha-2} < C|\theta - \theta'|^{2\alpha-2}.$$

Lemma 1.7.8. *Let*

$$I_n(k, \theta, \theta') := \int_{\mathbb{R}^3 \times \ldots \times \mathbb{R}^3} dx_0 \ldots dx_n f(x_0, \ldots, x_n) e^{ik[\theta' \cdot x_0 - \theta \cdot x_n + h(x_0, \ldots, x_n)]},$$

where

$$f(x_0, \ldots, x_n) = \frac{V(x_0) \cdots V(x_n)}{|x_0 - x_1||x_1 - x_2| \cdots |x_{n-1} - x_n|},$$

$$h(x_0, \ldots, x_n) = |x_0 - x_1| + |x_1 - x_2| + \cdots + |x_{n-1} - x_n|.$$

Then for $m = 0, 1, \ n = 2, 3, \ldots,$

$$\int_0^{\infty} d\alpha \, \alpha^m \int_{\mathbf{S}^2 \times \mathbf{S}^2} |\hat{I}_n(\alpha, \theta, \theta')|^2 < \infty,$$

where $\hat{I}$ *is the Fourier transform of* I.

Lemmas 1.7.4 and 1.7.5 imply that for $n = 0, 1$

$$\int_0^\infty d\alpha\, \alpha^n \|G'(\alpha)\|_{\mathbf{S}^2}^2 < \infty.$$

Now

$$
\begin{aligned}
\|\mathscr{G}\|_2^2 &= \int_0^\infty d\alpha \int_0^\infty d\beta \int_{\mathbf{S}^2 \times \mathbf{S}^2} d\theta\, d\theta'\, |G(\alpha + \beta, \theta, \theta')|^2 \\
&= \int_{\mathbf{S}^2 \times \mathbf{S}^2} d\theta\, d\theta' \int_0^\infty d\alpha\, \alpha\, |G(\alpha, \theta, \theta')|^2 = \int_0^\infty d\alpha\, \alpha \|G(\alpha)\|_{\mathbf{S}^2}^2
\end{aligned}
$$

by integration by parts. Therefore $\|\mathscr{G}'\|_2 < \infty$ and hence $\|\mathscr{G}'^2\|_2 \le \|\mathscr{G}'\|_2^2 < \infty$. By Lemma 1.7.6 $\|\mathscr{G}^{(0)}\mathscr{G}'\|_2 \le \|\mathscr{G}^{(0)}\| \, \|\mathscr{G}'\|_2 < \infty$ and consequently $\|\mathscr{G}^2\|_2 < \infty$, and $\mathscr{G}$ is bounded since both $\mathscr{G}^{0)}$ and $\mathscr{G}'$ are bounded. That $\mathscr{G}$ is self–adjoint follows from the reciprocity property of A. The proof that $\mathscr{G}^*$ has the same properties is exactly the same, except that A is replaced by $-\bar{A}$. This proves Lemma 1.5.14, provided we prove Lemmas 1.7.4 to 1.7.8.

Proof of Lemma 1.7.4. We find by direct computation that

$$
\begin{aligned}
k^2 \|A^{(R)}\|_{\mathbf{S}^2}^2 &= \int dx\, dy\, dx'\, dy'\, V^{1/2}(x)V^{1/2}(x')|V|^{1/2}(y)|V|^{1/2}(y') \\
&\quad \times [(\mathbb{1} - L)^{-1}L^2](x, y)\overline{[(\mathbb{1} - L)^{-1}L^2]}(x', y') \\
&\quad \times \frac{\sin k|x - x'| \sin k|y - y'|}{|x - x'|\,|y - y'|} \\
&\le \int dx\, dy\, dx'\, dy'\, \frac{|V(x)V(x')V(y)V(y')|}{|x - x'|\,|y - y'|} \\
&\quad \times |[(\mathbb{1} - L)^{-1}L^2](x, y)|\,|[(\mathbb{1} - L)^{-1}L^2](x', y')| \\
&\le \int dx\, dy\, \frac{|V(x)V(y)|}{|x - y|^2} \|(\mathbb{1} - L)^{-1}L^2\|_2^2 \\
&\le C \|L(k)^2\|_2^2
\end{aligned}
$$

by Schwarz's inequality. Therefore by Corollary 1.7.3, we have $\int dk\, k^2 \|A^{(R)}\|_{\mathbf{S}^2}^2 < \infty$ and consequently $\int d\alpha\, \|G^{(R)}\|_{\mathbf{S}^2}^2 < \infty$.

We must also examine the k–derivative of $A^{(R)}$. When the derivative acts on $\exp[ik(\theta' \cdot y - \theta \cdot x)]$ it simply brings down an additional factor of $|x|$ or $|y|$, which by the definition of $\mathscr{W}$ still leads to a finite result. When it acts on the function $k(\mathbb{1} - L)^{-1}$ we use Lemmas 1.1.3 and 1.2.2 and find that $\int dk \|\partial(kA^{(R)})/\partial k\|_{\mathbf{S}^2}^2 < \infty$. Both $[\mathbb{1} - L(k)]^{-1}$ and $\partial L/\partial k$ are uniformly bounded operator families and we therefore get $\int dk\, k^2\, \|\partial A^{(R)}/\partial k\|_{\mathbf{S}^2}^2 < \infty$. As a result we have $\int d\alpha\, \alpha^2 \|G^{(R)}\|_{\mathbf{S}^2}^2 < \infty$ and therefore also $\int d\alpha\, |\alpha|\, \|G^{(R)}\|_{\mathbf{S}^2}^2 < \infty$. $\qquad\square$

Proof of Lemma 1.7.5. The integrals in the left-hand side of the inequality to be proved are of the form that was estimated in Lemma 1.7.8. Therefore this result is a corollary of that lemma. $\qquad\square$

Proof of Lemma 1.7.6.

$$G^{(0)}(\alpha, \theta, \theta') = \frac{i}{(2\pi)^2} \int_{-\infty}^{\infty} dk\, k A^{(0)}(k, -\theta, \theta') e^{ik\alpha}$$

$$= \frac{1}{8\pi^2} \frac{\partial}{\partial \alpha} \int_{\mathbb{R}^3} dx\, V(x) \delta[\alpha + x \cdot (\theta + \theta')]$$

$$= \frac{1}{8\pi^2 |\theta + \theta'|^2} \int dx\, (\theta + \theta') \cdot \nabla V(x) \delta[\alpha + (\theta + \theta') \cdot x].$$

Therefore, by (iv) in the definition of $\mathscr{W}$

$$|G^{(0)}(\alpha, \theta, \theta')| \leq \frac{1}{8\pi^2 |\theta + \theta'|^2} \int_{\mathbb{R}^2} dx_\perp N\left[\left(|x_\perp|^2 + \frac{\alpha^2}{|\theta + \theta'|^2}\right)^{1/2}\right]$$

$$= \frac{1}{4\pi |\theta + \theta'|^2} F^2 \left(\frac{\alpha}{|\theta + \theta'|}\right),$$

where F is defined in (iv) of $\mathscr{W}$. Since F is monotone,

$$|G^{(0)}(\alpha, \theta.\theta')| \leq \frac{1}{4\pi} F(\alpha/2) \left(\frac{\alpha}{|\theta + \theta'|}\right)^\epsilon F\left(\frac{\alpha}{|\theta + \theta'|}\right) \alpha^{-\epsilon} |\theta + \theta'|^{\epsilon-2}$$

$$\leq CF(\alpha/2)\alpha^{-\epsilon}|\theta + \theta'|^{\epsilon-2}.$$

Now define

$$B(\alpha, \beta, \theta, \theta') := \int_0^\infty dt \int_{\mathbb{S}^2} d\theta''\, G^{(0)}(\alpha + t, \theta, \theta'') G^{(0)}(\beta + t, \theta'', \theta).$$

Then by Lemma 1.7.7

$$|B(\alpha, \beta, \theta, \theta')| \leq CF^{1/2}(\alpha/2) F^{1/2}(\beta/2) \alpha^{-\epsilon/2} \beta^{-\epsilon/2} \int_0^\infty dt\, t^{-\epsilon} F(t/2)$$

$$\times \int_{\mathbb{S}^2} d\theta'' |\theta + \theta''|^{\epsilon-2} |\theta' + \theta''|^{\epsilon-2}$$

$$\leq C|\theta + \theta'|^{2\epsilon-2} \alpha^{-\epsilon/2} \beta^{-\epsilon/2} F^{1/2}(\alpha/2) F^{1/2}(\beta/2)$$

$$\times \int_0^\infty dt\, t^{-\epsilon} F(t/2).$$

Therefore,

$$\|\mathscr{G}^{(0)2}\|_2^2 = \int_0^\infty d\alpha \int_0^\infty d\beta \int_{\mathbb{S}^2 \times \mathbb{S}^2} d\theta d\theta' \, |B(\alpha, \beta, \theta, \theta')|^2$$

$$\leq C\left[\int_0^\infty dt\, t^{-\epsilon} F(t/2)\right]^4 \int_{\mathbb{S}^2 \times \mathbb{S}^2} d\theta d\theta' \, |\theta + \theta'|^{4\epsilon-4}.$$

The last integral converges for $\epsilon > \frac{1}{2}$; the first converges for $\frac{1}{2} < \epsilon < \frac{3}{4}$ since the assumptions on N imply that $F(s) < Cs^{-1/4}$. Thus by the assumption on ϵ both converge.

Finally, the operator $\mathscr{G}^{(0)}$ is self-adjoint because of the reciprocity property of $A^{(0)}$. Therefore, $\|\mathscr{G}^{(0)} f\|^2 = (f, \mathscr{G}^{(0)2} f) \leq \|\mathscr{G}^{(0)2}\| \|f\|^2 \leq \|\mathscr{G}^{(0)}\|_2^2 \|f\|^2 \leq C\|f\|^2$ and $\mathscr{G}^{(0)}$ is bounded. □

Proof of Lemma 1.7.7. In the integration over the variable θ'' we use spherical polar coordinates with $\theta + \theta'$ as the z-axis and $\theta - \theta'$ as the x-axis, setting

$\theta'' \cdot (\theta + \theta') = |\theta + \theta'|u$, $\theta'' \cdot (\theta - \theta') = |\theta - \theta'|\sqrt{1 - u^2} \cos \varphi$, so that $|\theta - \theta''|^2 |\theta' - \theta''|^2 = (2 - |\theta + \theta'|u)^2 - |\theta - \theta'|(1 - u^2) \cos^2 \varphi$, and hence

$$I = \int_{-1}^{1} du \int_{0}^{2\pi} d\varphi \left[4 \left(1 - \tfrac{1}{2}|\theta + \theta'|u\right)^2 - |\theta - \theta'|(1 - u^2) \cos^2 \varphi \right]^{1/2\alpha - 1}.$$

It is readily seen that the integral

$$\int_{0}^{\pi} d\varphi \, |a^2 - \cos^2 \varphi|^{1/2\alpha - 1} = 2 \int_{0}^{1} dt \, (1 - t^2)^{-1/2} |a^2 - t^2|^{1/2\alpha - 1}$$

is $O(|a^2 - 1|^{1/2\alpha - 1/2})$ near $a^2 = 1$. Since it is $O(|a^2 - 1|^{1/2\alpha - 1})$ as $a^2 \to \infty$, $\exists C$ such that for all a

$$\int_{0}^{\pi} d\varphi \, |a^2 - \cos^2 \varphi|^{1/2\alpha - 1} < C \frac{|a^2 - 1|^{1/2\alpha - 1/2}}{(1 + |a^2 - 1|)^{1/2}}.$$

It follows that if $a^2 \geq b^2$ then

$$\int_{0}^{\pi} d\varphi \, (a^2 - b^2 \cos^2 \varphi)^{1/2\alpha - 1} \leq C(a^2 - b^2)^{1/2\alpha - 1/2} |a|^{-1}.$$

In the present case $a^2 = (2 - |\theta + \theta'|u)^2$, $a^2 - b^2 = (|\theta + \theta'| - 2u)^2$. Therefore

$$I \leq C \int_{-1}^{1} du \, | \, |\theta + \theta'| - 2u|^{\alpha - 1} |2 - |\theta + \theta'|u|^{-1}.$$

This integral converges for all θ and θ', except when $\theta = \theta'$, where one easily finds that it is $O(|\theta - \theta'|^{2\alpha - 2})$. ☐

Proof of Lemma 1.7.8. We have the Fourier transform

$$\widehat{I}_n(\alpha, \theta, \theta') = \frac{1}{2\pi} \int_{-\infty}^{\infty} dk \, ke^{ik\alpha} I_n(k, \theta, \theta')$$

$$= i \int dx_0 \cdots dx_n \, \delta \left[\alpha + \theta' \cdot x_0 - \theta \cdot x_n + h(x_0, \ldots, x_n)\right] \theta \cdot \nabla_0 f,$$

and hence

$$\int_{-\infty}^{\infty} d\alpha \, |\widehat{I}_n(\alpha, \theta, \theta')|^2$$

$$= \int dx_0 \cdots dx_n \, dx_0' \cdots dx_n' \, \delta \left[\theta' \cdot (x_0 - x_0') - \theta \cdot (x_n - x_n') + h(x_0, \ldots, x_n)\right]$$

$$-h(x_0', \ldots, x_n')] \theta' \cdot \nabla_0 f(x_0, \ldots, x_n) \theta' \cdot \nabla_0' f(x_0', \ldots, x_n').$$

The integral $\int d\alpha \, \alpha |\widehat{I}_n|^2$ has a factor of $[\theta' \cdot x_0' - \theta \cdot x_n' + h(x_0', \ldots, x_n')] \leq 2(|x_0'| + \cdots |x_n'|)$ in its integrand. Therefore for $m = 0, 1$

$$\int_{0}^{\infty} d\alpha \, \alpha^m \int d\theta \, d\theta' |\widehat{I}_n(\alpha, \theta, \theta')|^2$$

$$< C \int dx_0 \cdots dx_n' \, |\nabla_0 f(x_0, \ldots, x_n)| \, |\nabla_0' f(x_0', \ldots, x_n')| \frac{(|x_0'| + \ldots + |x_n'|)^m}{|x_0 - x_0'|}.$$

If $V \in \mathscr{W}$ the integrals over the variables $x_i, x_i', i = 1, \ldots, n$ all converge and are uniformly bounded. The subsequent integrals over x_0 and x_0' are of the form $\int dx \, dy \, |w_1(x)w_2(y)|/|x - y|$, where w_1 and w_2 are either V or ∇V. These integrals all converge if $V \in \mathscr{W}$. Note that the assumption (iv) in the definition of $\mathscr{W}$ implies that $\int_{-\infty}^{\infty} dt \, t^2 N(t) < \infty$. ☐

1.8 Notes

Further details on the direct scattering problem can be found in [AJS77], [RS79], and [Ne82c].

1.2 Lemma 1.2.1 was first proved in [Ne74c]. The Birman-Schwinger bound (1.20) comes from [Bi61] and [Sc61]. Exceptional points of the first and second kind were first distinguished in [Ne77a]. The character functions of bound states, which were introduced in [Ne80], reduce to the spherical harmonics for central potentials. The notion of *normal* eigenvalues and *normal* potentials is new. It is not known under what conditions a potential is normal. The method given for recovering the character spaces from the scattering amplitude is new.

1.3 The linear independence of the functions v' defined in (1.50) is an interesting open problem.

1.4 The results in (1.4.2) were given in [Ne80].

1.5 Levinson's theorem was generalized to three dimensions by [Dr76, 78a, 78b, and Ne77]. Lemma 1.5.11 is an important improvement over an analogous lemma in [Ne80]. It was proved in the preprint [We88], which I received in November, 1988. Actually, the version given there is somewhat more general than that stated in Lemma 1.5.11.

1.7 The proofs of Lemmas 1.4.2 and 1.5.14 are improved versions of those given in [Ne80, 81, and 83]. The proof of Lemma 1.5.11 is a somewhat altered and more detailed version of that in [We88]; see also [We85 and 89] and [Ku78].

2. The Inverse Problem

2.1 Introduction and Uniqueness

We now turn to the inverse problem, in which it is assumed that the scattering amplitude is given as a function, $\mathbb{R} \times S^2 \times S^2 \mapsto \mathbb{C}$, and the aim is to infer the underlying potential in the Schrödinger equation (1.1). It should be noted that from a physical point of view the experimental data are, at best, given by the differential scattering cross section $|A|^2$ rather than the complex scattering amplitude itself. There is, therefore, a first step in the inverse problem that will not be discussed here: to infer the scattering amplitude from a knowledge of its modulus. For this step we refer to the literature (see, for example, [Ne82c, Section 20.2]). For our purposes the data will be considered to be the scattering amplitude $A(k, \theta, \theta')$.

The first issue in the inversion problem to be discussed here is the question of the *uniqueness* of its solution. This can be dealt with rather easily in the present case. We saw in Section 1.5 that for large $|k|$, $k \in \mathbb{R}$, with the momentum transfer $\tau = k(\theta' - \theta)$ fixed, the scattering amplitude approaches the Born approximation, as in (1.85). Now the Born approximation (1.83),

$$B(\tau) = -\frac{1}{4\pi} \int_{\mathbb{R}^3} dx \, V(x) e^{ik\tau \cdot x}$$

is, apart from a constant factor, the Fourier transform of the potential. At a fixed value of k, the momentum transfer is necessarily bounded by $|\tau| \leq 2|k|$, and knowledge of the Born approximation does not imply knowledge of the entire Fourier transform of the potential. However, as $|k| \to \infty$, this limitation becomes ineffective and knowledge of the large-$|k|$ limit of $A(k, \theta, \theta')$ with fixed τ implies knowledge of the entire Fourier transform of the potential, and hence knowledge of the potential.

Lemma 2.1.1. *If* $\lim_{k \to \pm\infty, k(\theta'-\theta)=\tau} A(k, \theta, \theta')$ *is known then the underlying potential* $V(x)$ *is uniquely determined for almost all* $x \in \mathbb{R}^3$.

In fact, since for any fixed τ, as $|k| \to \infty$, θ' approaches θ, the knowledge of V obtained comes essentially all from the forward diffraction peak.

Since the large-$|k|$ limit of the scattering amplitude directly leads to the Fourier transform of the potential, one may consider the inverse problem solved: To determine the potential one need only observe the scattering amplitude at large $|k|$ and invert the Fourier transform. (A variant of this large-$|k|$ inversion due to Y. Saitō will be briefly discussed in Section 2.9.) However, there are two relevant remarks that make this solution not very satisfactory. The first is the observation that reliance on the short-wavelength limit is inherently inconvenient from an experimental point of view. All the relevant information being contained in the forward diffraction peak, the results are obviously sensitive to small errors in the data very close to the forward direction, which are usually hard to obtain.

The second remark is the observation that if the potential is obtained from the large-$|k|$ limit of the scattering amplitude, changing a given scattering amplitude at any finite value of k will not alter the result. Therefore this procedure cannot lead to any characterization of admissible scattering amplitudes, i.e., it cannot tell us what functions $A(k, \theta, \theta')$ are actually obtainable as scattering amplitudes from a Schrödinger equation (1.1). That this question is not at all trivial can be seen from the simple fact that A is a function on a five-dimensional manifold (the five real variables consist of k and two angles each for θ and $\theta' \in S^2$), while V is a function on three-dimensional Euclidian space. It is therefore to be expected (and will be confirmed later on) that the functions A admissible as scattering amplitudes constitute an extremely restricted set. Or put another way: The data contained in $A(k, \theta, \theta')$ are highly redundant. One of the aims of solving the inverse scattering problem is to get information on this redundancy, and the large-$|k|$ limit cannot provide that.

Remark. It should be noted that the uniqueness statement of Lemma 2.1.1 has no analogue in the inverse scattering problem in one dimension, nor does it have an analogue in the inverse problem in three dimensions for central potentials if a single phase shift is given for all $k \in \mathbb{R}$. In both of these cases the bound states are independent of the scattering amplitude (except for the Levinson theorem), while here they are not.[1]

2.2 A First Approach

The most direct method of attacking the inverse scattering problem is to use the relation (1.61) established earlier:

$$\psi^+(-k, \theta, x) = \int_{S^2} d\theta' \, S(-k, -\theta', \theta) \psi^+(k, \theta', x).$$

Let us rewrite this in terms of the function ζ defined in (1.52) and use the reciprocity relation (1.70). The result is

$$\zeta(-k, \theta, x) = \int_{S^2} d\theta' \, S_x(-k, -\theta, \theta') \zeta(k, \theta', x), \tag{2.1}$$

where S_x is the S matrix of the potential V shifted by x, $V_x(\cdot) = V(\cdot + x)$, as in (1.81).

For simplicity consider the case without bound states. We are looking for a function that satisfies (2.1) and that has the properties established in Section 1.1: For each fixed $x \in \mathbb{R}^3$, ζ should be the boundary value of an analytic function that is holomorphic in $\mathbb{C}^+$ and that approaches unity as $|k| \to \infty$. The search for such a function is a generalization of a classical Riemann-Hilbert problem in which the *symbol* S_x is an integral kernel. It can be converted into an integral equation by formulating the requirements on ζ by means of a Hilbert transform.

---

[1] This independence of the bound states from the phase shifts was, thirty five years ago, considered as evidence against the soundness of Heisenberg's aim of replacing the Hamiltonian by the S matrix as a basic dynamical tool of quantum mechanics. That argument can now be seen to have been somewhat rash.

The analyticity of $\zeta - 1$ in $\mathbb{C}^+$ together with the fact that $\zeta - 1 \to 0$ at infinity (Lemma 1.4.2) are equivalent to the statement that ζ satisfy the "dispersion relation"

$$\zeta(k, \theta, x) - 1 = \frac{1}{2\pi i} \int_{-\infty}^{\infty} dk' \frac{\zeta(k', \theta, x) - 1}{k' - k - i\epsilon}$$

in the limit as $\epsilon \downarrow 0$. Inserting (2.1) in this equation we obtain

$$\zeta(k, \theta, x) = 1 - \frac{1}{2\pi i} \int_{-\infty}^{\infty} dk' \frac{\zeta(k', \theta, x) - 1}{k' + k + i\epsilon}$$

$$+ \frac{1}{4\pi^2} \int_{-\infty}^{\infty} \frac{dk' \, k'}{k' + k + i\epsilon} \int_{S^2} d\theta' A_x(-k', -\theta, \theta') \zeta(k', \theta', x).$$

The second term on the right-hand side vanishes because of the analyticity of ζ and we obtain

$$\zeta(k, \theta, x)$$

$$= 1 + \frac{1}{4\pi^2} \int_{-\infty}^{\infty} \frac{dk' \, k'}{k' + k + i\epsilon} \int_{S^2} d\theta' A(-k', -\theta, \theta') \zeta(k', \theta', x) e^{ik' x \cdot (\theta + \theta')}. \quad (2.2)$$

This may be regarded as a (singular) integral equation for ζ. If it can be solved then ψ^+ is obtained from (1.52) and the potential V may be recovered from the Schrödinger equation (1.1). Alternatively, the potential may be computed by inserting (2.2) in the differential equation (1.53) and letting $k \to \infty$. This results in the following representation for the potential:

$$V(x) = \frac{i}{2\pi^2} \theta \cdot \nabla \int_{-\infty}^{\infty} dk \, k \int_{S^2} d\theta' A(-k, -\theta, \theta') \psi^+(k, \theta', x) e^{ik\theta \cdot x}. \quad (2.3)$$

It is not difficult to generalize this method to the case with bound states. However, the singular nature of (2.2) makes it less useful, both from a practical point of view and for purposes of studying the existence and uniqueness of its solutions. We shall therefore not pursue this avenue further. Nevertheless, equation (2.2) will be of use at certain points.

We shall instead study the solution of the Riemann-Hilbert problem that arises from the equation (2.1) by subjecting it to Fourier transformation. It will be very advantageous to use the operator language already introduced in Section 1.1 and regard $A_x(k, \theta, \theta')$ and $S_x(k, \theta, \theta')$ as kernels of operator families $A_x(k)$ and $S_x(k)$. Similarly, we regard $\zeta(k, \cdot, x)$ as a family of members $\zeta_x(k)$ of a Banach space. Furthermore we use the operator Q defined just above Definition 1.5.15, $(Qf)(\theta) := f(-\theta)$, and also use the notation

$$f^*(k) := f(-k).$$

Note that the $*$-notation refers only to the k-dependence of all functions and Q refers only to their θ-dependence. In this notation (2.1) reads simply

$$\zeta_x^* = QS_x^* \zeta_x, \quad (2.4)$$

in which x is to be regarded as a parameter.

2.3 The Riemann-Hilbert Problem

2.3.1 No Bound States

Before formulating precisely the Riemann-Hilbert problem that has to be solved we define $\hat{1}$ to be the function $S^2 \mapsto \mathbb{C}$, which is identically equal to unity: $\hat{1}(\theta) \equiv 1$. Here then is the Riemann-Hilbert problem:

Problem $H_0^1(S_x)$: *Let*

$$\Omega(k) := QS_x(-k) = Q + \frac{k}{2\pi i}QA_x(-k), \tag{2.5}$$

where $A_x(k)$ is the operator whose kernel is given by $A_x(k,\theta,\theta') := A(k,\theta,\theta')e^{ikx\cdot(\theta-\theta')}$, $\theta,\theta',x \in \mathbb{R}^3$, $|\theta| = |\theta'| = 1$, and $A(k,\theta,\theta') \in \mathscr{A}$, as defined in Definition 1.5.15, and Q is defined just before that definition. Find a function f, $\mathbb{R} \times S^2 \mapsto \mathbb{C}$, which is such that

(i) $f - \hat{1} \in L^2(\mathbb{R} \times S^2)$;

(ii) $f(k)$, $\mathbb{R} \mapsto L^2(S^2)$, is the boundary value of an analytic function holomorphic in $\mathbb{C}^+$ such that $\lim_{|k|\to\infty} \|f(k) - \hat{1}\| = 0$; [Here $\|\cdot\|$ is the norm in $L^2(S^2)$.]

(iii) on $\mathbb{R}$ f satisfies the equation

$$f^* = \Omega f, \tag{2.6}$$

where $f^*(k) = f(-k)$.

It follows from Lemma 1.4.2 that if ζ is connected to ψ^+ as in (1.52), where ψ^+ solves the Lippmann-Schwinger equation (1.3) with a potential in $\mathscr{W}$ that causes no bound states and A in (2.5) is the corresponding scattering amplitude as in (1.2), then ζ solves H_0^1. The question is (a) is it the only solution? and (b) if A is given, how do we find f? Further, if $A \in \mathscr{A}_0$ is given and we don't know that an underlying potential exists, does H_0^1 have a solution?

In order to solve H_0^1 we rewrite (2.6) by adding and subtracting $\hat{1}$:

$$f^* - \hat{1} = (\Omega - Q)(f - \hat{1}) + (\Omega - Q)\hat{1} + Q(f - \hat{1}), \tag{2.7}$$

and subjecting it to Fourier transformation. We define the Fourier transforms

$$G(\alpha) := \frac{1}{2\pi} \int_{-\infty}^{\infty} dk\, e^{ik\alpha}[\Omega(k) - Q] \tag{2.8}$$

[as in (1.86) for $x = 0$] and

$$\eta(\alpha) := \frac{1}{2\pi} \int_{-\infty}^{\infty} dk\, e^{-ik\alpha}[f(k) - \hat{1}]. \tag{2.9}$$

According to (ii) in H_0^1 the Fourier transform of $f - \hat{1}$ is well-defined in $L^2(\mathbb{R} \times S^2)$ as a limit in the mean. Since the unitarity of S implies that $\|S(k)\| = 1$, where $\|\cdot\|$ is the norm in $L^2(S^2)$, $(\Omega - \mathbb{1})(f - \hat{1})$ is also in $L^2(\mathbb{R} \times S^2)$, and hence its Fourier transform exists in L^2. Finally, by Lemmas 1.5.11 and 1.5.13 the Fourier transform of $kA(k)$ exists in the L^2-sense and is in $\mathscr{L}^2$. Therefore the functions $G(\alpha)$ and $\eta(\alpha)$ are well-defined and Fourier transformation of (2.7) leads to the equation

$$\eta(\alpha) = Q\eta(-\alpha) + g(\alpha) + \int_{-\infty}^{\infty} d\beta \, G(\alpha + \beta)\eta(\beta), \tag{2.10}$$

where

$$g(\alpha) = G(\alpha)\hat{1}, \tag{2.11}$$

which means explicitly,

$$g(\alpha, \theta, x) = \frac{i}{4\pi^2} \int_{-\infty}^{\infty} dk \, k \int_{S^2} d\theta' \, A(k, -\theta, \theta') e^{-ik[\alpha + x \cdot (\theta + \theta')]}. \tag{2.12}$$

Now the requirement (ii) on f implies that for $\alpha < 0$, $\eta(\alpha) = 0$. Therefore it is convenient to rewrite (2.10) separately for $\alpha > 0$,

$$\eta(\alpha) = g(\alpha) + \int_{0}^{\infty} d\beta \, G(\alpha + \beta)\eta(\beta), \tag{2.13}$$

and for $\alpha < 0$,

$$Q\eta(-\alpha) = -g(\alpha) - \int_{0}^{\infty} d\beta \, G(\alpha + \beta)\eta(\beta),$$

or, writing $\alpha \to -\alpha$, for $\alpha > 0$,

$$\eta(\alpha) = -Qg(-\alpha) - \int_{0}^{\infty} d\beta \, QG(\beta - \alpha)\eta(\beta). \tag{2.14}$$

Equations (2.13) and (2.14) for $\alpha > 0$ are equivalent to (2.6) together with (ii) in H_0^1 by Fourier transformation (2.9). That is, if $\eta \in L^2(\mathbb{R}_+ \times S^2)$ satisfies both (2.13) and (2.14) on $\mathbb{R}_+$ then

$$f(k) = \hat{1} + \int_{0}^{\infty} d\alpha \, e^{ik\alpha}\eta(\alpha) \tag{2.15}$$

solves H_0^1.

To simplify our writing let us define operators $\mathscr{G}$ and $\mathscr{H}$ by the kernels

$$\mathscr{G}(\alpha, \theta; \beta, \theta') \;\; := \;\; G(\alpha + \beta; \theta, \theta'), \tag{2.16}$$
$$\mathscr{H}(\alpha, \theta; \beta, \theta') \;\; := \;\; G(\alpha - \beta; \theta, \theta'), \tag{2.17}$$

where $\alpha, \beta \in \mathbb{R}_+$, $\theta, \theta' \in S^2$, and similarly,

$$\mathscr{G}^*(\alpha, \theta; \beta, \theta') \;\; := \;\; G(-\alpha - \beta; \theta, \theta'), \tag{2.18}$$
$$\mathscr{H}^*(\alpha, \theta; \beta, \theta') \;\; := \;\; G(\beta - \alpha; \theta, \theta'). \tag{2.19}$$

[$\mathscr{G}$ and $\mathscr{G}^*$ had already been defined in (1.87) and (1.88) for $x = 0$.] It is easily seen that the operator $\mathscr{G}$ is self-adjoint while $\mathscr{H}^*$ is the adjoint of $\mathscr{H}$. Then (2.13) and (2.14) can be written in operator notation, on $\mathbb{R}_+ \times S^2$

$$\eta \;\; = \;\; g + \mathscr{G}\eta, \tag{2.20}$$
$$\eta \;\; = \;\; -Qg^* - Q\mathscr{H}^*\eta. \tag{2.21}$$

Equation (2.13) or (2.20) is a generalization of the Marchenko equation to three dimensions. As we shall see, its kernel has desirable mathematical properties.

Equation (2.14) or (2.21), on the other hand, is a Wiener-Hopf equation and it is difficult to solve directly. Intuitively, the reason for the difference in the behavior of the kernels $\mathcal{G}$ and $\mathcal{H}^*$ is the fact that $\mathcal{G}$ is a function of $\alpha + \beta$ while $\mathcal{H}^*$ is a function of $\alpha - \beta$. As α and β tend to infinity there can be no cancellation between them in $\mathcal{G}$, but there can be in $\mathcal{H}^*$. Thus, as $\alpha, \beta \to \infty$ the argument of the function $\mathcal{G}$ tends to infinity, but that of $\mathcal{H}^*$ does not necessarily. The crux of the Marchenko method is to solve only (2.13) or (2.20) and to avoid having to solve (2.14) or (2.21). Under certain conditions it will turn out, as we shall see, that a solution of (2.20) alone will, in fact, solve (2.21) as well.

It may be helpful to write out the generalized Marchenko equation (2.13) explicitly with all its variables visible. This is done most simply by defining the function

$$\widehat{\eta}(\alpha, \theta, x) := \eta(\alpha - \theta \cdot x, \theta, x). \tag{2.22}$$

Then equations (2.8), (2.5), and (1.81) show that equation (2.13) reads, for $\alpha > \theta \cdot x$

$$\widehat{\eta}(\alpha, \theta, x) = \int_{S^2} d\theta' \, G(\alpha + x \cdot \theta', \theta, \theta')$$
$$+ \int_{S^2} d\theta' \int_{x \cdot \theta'}^{\infty} d\beta \, G(\alpha + \beta, \theta, \theta') \widehat{\eta}(\beta, \theta', x), \tag{2.23}$$

where

$$G(\alpha, \theta, \theta') = \frac{i}{(2\pi)^2} \int_{-\infty}^{\infty} dk \, k A(k, -\theta, \theta') e^{-ik\alpha} \tag{2.24}$$

and the scattering solution of (1.1) is connected to $\widehat{\eta}$ by

$$\psi^+(k, \theta, x) = e^{ik\theta \cdot x} + \int_{x \cdot \theta}^{\infty} d\alpha \, \widehat{\eta}(\alpha, \theta, x) e^{ik\alpha}. \tag{2.25}$$

Theorem 2.3.1. *If the scattering amplitude A is in the class $\mathcal{A}$ then $\mathcal{G}$ and $\mathcal{G}^*$ are operators $L^2(\mathbb{R}_+ \times S^2) \mapsto L^2(\mathbb{R}_+ \times S^2)$ that are self-adjoint and compact, and $\|\mathcal{G}\| \leq 1$, $\|\mathcal{G}^*\| \leq 1$.*

Proof. Definition 1.5.15, item *(viii)*, says that $\mathcal{G}$ and $\mathcal{G}^*$ are bounded and self-adjoint, and that $\mathcal{G}^2$ and $\mathcal{G}^{*2}$ are Hilbert-Schmidt. Therefore they are compact. Before we prove that $\|\mathcal{G}\| \leq 1$ and $\|\mathcal{G}^*\| \leq 1$ let us translate the unitarity of $S(k)$ into a statement about $\mathcal{G}$ and $\mathcal{H}$. This is a matter of straight-forward Fourier transformation.

Lemma 2.3.2. *If for each $k \in \mathbb{R}$, $A(k)$ has properties (ii), (iii), and (iv) in Definition 1.5.15 the following equations hold:*

$$(Q + \mathcal{H})(Q + \mathcal{H}^*) = \mathbb{1} - \mathcal{G}^2, \tag{2.26}$$

$$(Q + \mathcal{H}^*)\mathcal{G} + \mathcal{G}^*(Q + \mathcal{H}^*) = 0, \tag{2.27}$$

*as well as those obtained from them by the *-operation:*

$$(Q + \mathcal{H}^*)(Q + \mathcal{H}) = \mathbb{1} - \mathcal{G}^{*2}, \tag{2.28}$$

$$(Q + \mathcal{H})\mathcal{G}^* + \mathcal{G}(Q + \mathcal{H}) = 0, \tag{2.29}$$

Furthermore, the function g *in the generalized Marchenko equation (2.13) satisfies the following equations:*

$$(\mathbb{1} + \mathcal{G})g = -(Q + \mathcal{H})g^*, \tag{2.30}$$

$$(\mathbb{1} + \mathcal{G}^*)g^* = -(Q + \mathcal{H}^*)g. \tag{2.31}$$

Because $\mathcal{H}^* = \mathcal{H}^\dagger$ and $Q = Q^\dagger$ equations (2.26) and (2.28) imply that $\mathbb{1} - \mathcal{G}^2$ and $\mathbb{1} - \mathcal{G}^{*2}$ are non-negative. This proves the theorem.   □

It follows from Theorem 2.3.1 that the kernel $\mathcal{G}$ of the generalized Marchenko equation has a pure point spectrum that lies in the interval $[-1, 1]$. If $V \in \mathcal{W}$ and it produces no bound states then, by Theorem 1.5.16, $A_x \in \mathcal{A}_0$ and we know that the corresponding function ζ satisfies the Riemann-Hilbert problem $H_0^1(S_x)$. Therefore in that case we are assured that a solution of (2.13) exists. However, if we start with an arbitrarily given function $A \in \mathcal{A}$ then we do not, at this point, know whether the generalized Marchenko equation (2.13) has a solution or not. We first observe that the assumption that $kA(k) \in \mathcal{L}^2$ assures that $g \in L^2(\mathbb{R}_+ \times S^2)$. Therefore, if $\mathcal{G}$ does not have the eigenvalue 1, then (2.13) has a solution in $L^2(\mathbb{R}_+ \times S^2)$. What happens if $\mathcal{G}$ has the eigenvalue 1? The answer is that even in that case (2.13) has a solution.

Lemma 2.3.3. *If* g *and* $\mathcal{G}$ *are defined in terms of A as in (2.12), (2.8),(2.5), and (2.16) and if $A \in \mathcal{A}$ then the generalized Marchenko equation (2.13) always has a solution in $L^2(\mathbb{R}_+ \times S^2)$.*

Proof. In order to prove the lemma we have to prove the following: If ξ satisfies the homogeneous form of the generalized Marchenko equation (2.13) then $\langle \xi, g \rangle_+ = 0$, where $\langle \cdot, \cdot \rangle_+$ is the inner product on $\mathbb{R}_+ \times S^2$. Suppose that $(\mathcal{G} - \mathbb{1})\xi = 0$. Then, using the self-adjointness of $\mathcal{G}$, equation (2.30), the fact that $\mathcal{H}^* = \mathcal{H}^\dagger$, and equation (2.26), we have $2\langle g, \xi \rangle_+ = \langle g, (\mathbb{1} + \mathcal{G})\xi \rangle_+ = \langle (\mathbb{1} + \mathcal{G})g, \xi \rangle_+ = -\langle (Q + \mathcal{H})g^*, \xi \rangle_+ = -\langle g^*, (Q + \mathcal{H}^*)\xi \rangle_+ = 0$.   □

Thus the generalized Marchenko equation is always solvable if $A \in \mathcal{A}$, though not necessarily uniquely. The next question is: Does a solution of (2.13) in L^2 necessarily lead to a solution of the Riemann-Hilbert problem ? The answer is this:

Lemma 2.3.4. *If $\mathcal{G}^*$ does not have the eigenvalue -1 then the right-handed Fourier transform as in (2.15) of any solution of the generalized Marchenko equation (2.20) solves the Riemann-Hilbert problem $H_0^1(S_x)$.*

Proof. We prove this by proving that if $\mathcal{G}^*$ does not have the eigenvalue -1 then any solution of (2.20) also solves (2.21). Suppose that η solves (2.20). Define

$$\xi := (Q + \mathcal{H}^*)\eta + g^*.$$

Then, by (2.31), (2.27), and (2.13)

$$\begin{aligned}
(\mathbb{1} + \mathcal{G}^*)\xi &= (\mathbb{1} + \mathcal{G}^*)g^* + (Q + \mathcal{H}^*)\eta + \mathcal{G}^*(Q + \mathcal{H}^*)\eta \\
&= -(Q + \mathcal{H}^*)g + (Q + \mathcal{H}^*)\eta - (Q + \mathcal{H}^*)\mathcal{G}\eta \\
&= (Q + \mathcal{H}^*)(\eta - g - \mathcal{G}\eta) = 0.
\end{aligned}$$

Therefore, if -1 is not in the spectrum of $\mathcal{G}^*$, then $\xi = 0$ and hence η satisfies (2.21). Since the two equations (2.20) and (2.21) are equivalent to equation (2.6) together with the requisite analyticity and asymptotics, the lemma is proved. $\square$

We also note that if $\mathcal{G}$ has the eigenvalue 1, i.e., the homogeneous form of the generalized Marchenko equation (2.20) has a nontrivial solution σ, then (2.26) together with $\mathcal{H}^* = \mathcal{H}^\dagger$ implies that $(Q + \mathcal{H}^*)\sigma = 0$ and hence σ solves the homogeneous version of (2.21) too. But the homogeneous versions of (2.20) and (2.21) are, by right-handed Fourier transformation

$$f_0(k) = \int_0^\infty d\alpha\, e^{ik\alpha}\sigma(\alpha), \tag{2.32}$$

equivalent to the homogeneous Riemann-Hilbert problem given as follows:
Problem $H_0^0(S_x)$: *Let*

$$\Omega(k) := QS_x(-k) = Q + \frac{k}{2\pi i}QA_x(-k),$$

where $A_x(k)$ is the operator whose kernel is given by $A_x(k, \theta, \theta') = A(k, \theta, \theta')e^{ikx\cdot(\theta - \theta')}$, $\theta, \theta' \in \mathbb{R}^3$, $|\theta| = |\theta'| = 1$, $x \in \mathbb{R}^3$, and $A(k, \theta, \theta') \in \mathcal{A}$, as defined in Definition 1.5.15, and Q is defined just before that definition. Find a function f, $\mathbb{R} \times S^2 \mapsto \mathbb{C}$, which is such that

(i) $f \in L^2(\mathbb{R} \times S^2)$;

(ii) $f(k)$, $\mathbb{R} \mapsto L^2(S^2)$, is the boundary value of an analytic function holomorphic in $\mathbb{C}^+$ such that $\lim_{|k|\to\infty} \|f(k)\| = 0$; [Here $\|\cdot\|$ is the norm in $L^2(S^2)$.]

(iii) on $\mathbb{R}$ f satisfies the equation

$$f^* = \Omega f.$$

Of course, if $H_0^0(S_x)$ has a nontrivial solution, $H_0^1(S_x)$ does not have a unique solution.

In the same manner as above we find that if $\mathcal{G}$ has the eigenvalue -1 and σ is the corresponding eigenfunction then the Fourier transform (2.32) solves the problem $H_0^0(-S_x)$.

Lemma 2.3.5. *If and only if $H_0^0(\pm S_x)$ has a nontrivial solution, $\mathcal{G}$ has the eigenvalue ± 1 [where $\mathcal{G}$ is related to S_x by (2.8), (2.16), and (2.5)]. The eigenfunction σ of $\mathcal{G}$ with the eigenvalue ± 1 is related to the solution of H_0^0 as in (2.32).*

As a result we have proved the following.

Theorem 2.3.6. *If $\mathcal{G}^*$ does not have the eigenvalue -1 then the Riemann-Hilbert problem $H_0^1(S_x)$ has a solution and each such solution is obtained by right-handed Fourier transformation (2.15) from a solution of the generalized Marchenko equation (2.13). This solution is unique if and only if $\mathcal{G}$ does not have the eigenvalue 1.*

Now it should be noted that the operator $-\mathcal{G}^*$ is related to $\bar{S}_x$ in exactly the same manner as $\mathcal{G}$ is related to S_x. We therefore have the folowing situation:

(i) If neither $\mathcal{G}$ nor $-\mathcal{G}^*$ has the eigenvalue 1 then both $H_0^1(S_x)$ and $H_0^1(\bar{S}_x)$ have unique solutions.

(ii) If $\mathcal{G}$ has the eigenvalue 1 but $-\mathcal{G}^*$ does not, then $H_0^1(S_x)$ has more than one solution.

(iii) If $-\mathcal{G}^*$ has the eigenvalue 1 but $\mathcal{G}$ does not, then $H_0^1(\overline{S}_x)$ has more than one solution.

As far as solving the Riemann-Hilbert problem $H_0^1(S_x)$ is concerned, that has now been reduced to solving the Fredholm equation (2.13), whose kernel $\mathcal{G}$ is real (that follows from item (iii) in the definition of $\mathscr{A}$) and such that $\|\mathcal{G}\| \le 1$ if $A \in \mathscr{A}$. If, in fact, $\|\mathcal{G}\| < 1$ then the generalized Marchenko equation (2.13) not only has a unique solution, but that solution can be computed by iteration. (The Neumann series converges.) If furthermore $\|\mathcal{G}^*\| < 1$ ($\|\mathcal{G}^*\| \le 1$ is guaranteed if $A \in \mathscr{A}$) then the right-hand Fourier transform as in (2.15) of the solution in fact solves the Riemann-Hilbert problem $H_0^1(S_x)$.

Can it be guaranteed that the solution of the Riemann-Hilbert problem is not only in L^2 but in L^1, so that its Fourier transform is bounded? There is the following result.

Theorem 2.3.7. Let G be defined as in (2.8) in terms of the scattering amplitude. Suppose that $\|G(\cdot)\|(1 + \sqrt{\cdot}) \in L^1(\mathbb{R}_+)$, where $\|\cdot\|$ is either the operator norm or the Hilbert-Schmidt norm on $L^2(S^2)$. Then the right-hand Fourier transform as in (2.15) of the solution η of the generalized Marchenko equation is a uniformly bounded function of k.

Proof. Let $f \in L^2(\mathbb{R}_+ \times S^2)$; then by Schwarz's inequality and integration by parts

$$\int_0^\infty d\alpha \Big\| \int_0^\infty d\beta\, G(\alpha + \beta) f(\beta) \Big\|_2$$
$$\le \int_0^\infty d\alpha \int_0^\infty d\beta\, \|G(\alpha + \beta)\| \, \|f(\beta)\|_2$$
$$= \int_0^\infty d\beta \int_\beta^\infty d\alpha\, \|G(\alpha)\| \, \|f(\beta)\|_2$$
$$\le \Big[\int_0^\infty d\beta\, \|f(\beta)\|_2^2 \Big]^{1/2} \Big[\int_0^\infty d\beta \Big[\int_\beta^\infty d\alpha \|G(\alpha)\| \Big]^2 \Big]^{1/2}$$
$$= \|f\| \Big[2 \int_0^\infty d\beta\, \beta \|G(\beta)\| \int_\beta^\infty d\alpha\, \|G(\alpha)\| \Big]^{1/2}$$
$$\le \|f\| \Big[2 \int_0^\infty d\beta\, \beta^{1/2} \|G(\beta)\| \int_\beta^\infty d\alpha\, \alpha^{1/2} \|G(\alpha)\| \Big]^{1/2}$$
$$= \sqrt{2} \|f\| \int_0^\infty d\alpha\, \alpha^{1/2} \|G(\alpha)\|.$$

Here $\|G\|$ can be either the operator norm or the Hilbert-Schmidt norm. Thus, if $f \in L^2(\mathbb{R}_+ \times S^2)$ then $\|\mathcal{G}f\|_2 \in L^1(\mathbb{R}_+)$. Now take η, an L^2-solution of the generalized Marchenko equation (2.20), $\eta = g + \mathcal{G}\eta$. Then $\|\mathcal{G}\eta\|_2 \in L^1(\mathbb{R}_+)$, and $g = G\hat{1}$, $\|g\|_2 \le \|G\|$, and hence $\|g\|_2 \in L^1(\mathbb{R}_+)$ by one of the hypotheses. Therefore $\|\eta\|_2 \in L^1$ and its Fourier transform is uniformly bounded in k. □

2.3.2 Bound States

If there are bound states with eigenvalues $-\kappa_m^2$, the function ζ must still satisfy equation (2.4) but it is meromorphic in $\mathbb{C}^+$ with a finite number of simple poles at the points $i\kappa_m$. The first task is to find the needed bound-state data from the scattering amplitude.

Since the forward scattering amplitude $A(k, \theta, \theta)$ has an analytic continuation into $\mathbb{C}^+$ with simple poles at $i\kappa_m$, the numbers κ_m can be obtained from $A(k, \theta, \theta)$. This can be accomplished, for example, by Fourier transformation. The function $A(k, \theta, \theta)$ approaches a constant as $k \to \pm\infty$ [see (1.85)]. Therefore for $t > 0$

$$\frac{1}{2\pi i} \int_{-\infty}^{\infty} dk\, [A(k, \theta, \theta) - c]e^{ikt} = \frac{1}{2\pi t} \int_{-\infty}^{\infty} dk\, e^{ikt} \frac{\partial A(k, \theta, \theta)}{\partial k}$$

$$= \sum_m \operatorname{Res}_{i\kappa_m} A(k, \theta, \theta)e^{-\kappa_m t},$$

from which both the positive numbers κ_m and the residues of $A(k, \theta, \theta)$ can be recovered. Similarly we may obtain the various angle-derivatives defined by (1.66). It may therefore be assumed that the functions $A_{\kappa\varsigma}^{mn}(\theta)$ are known. If $A \in \mathscr{A}_n$ then by assumption (viii) in the definition of $\mathscr{A}_n$, the poles of the forward scattering amplitude are all normal, as defined by Definition 1.5.4, and the assumption that $n = \sum_m N_m'$ guarantees that for each m $N_m = N_m'$. That is because, if an underlying potential $V \in \mathscr{W}$ exists, then the generalized Levinson theorem together with assumption (viii) implies that $\sum_m N_m = \sum_m N_m'$. Lemma 1.5.5 then implies that for each m $N_m = N_m'$, and the eigenvalues are normal. Thus we know the degeneracy of each eigenvalue. Furthermore we may determine, by means of Lemma 1.5.7, the space $\mathscr{H}_\kappa$ spanned by the characters.

Lemma 2.3.8. *If $A \in \mathscr{A}_n$ and there exists an underlying potential in $\mathscr{W}$ then all the eigenvalues, their degeneracies, and their character spaces are uniquely determined by $A(k, \theta, \theta')$.*

We now turn to the Riemann-Hilbert problem that has to be solved when there are bound states.

Problem $H_\sigma^1(S_x)$: *Let $\Omega(k)$ be defined in terms of the given distribution kernel S_x with $A(k, \theta, \theta') \in \mathscr{A}$ as in (2.5), $\mathscr{A}$ being defined by Definition 1.5.15. Let κ_m be a given set of positive numbers and $\mathscr{H}_m$ a given set of finite-dimensional subspaces of $L^2(\mathbb{S}^2)$ in one-to-one correspondence with them, the sum of whose dimensions equals n_σ. We denote the set of positive numbers κ_m together with the corresponding spaces $\mathscr{H}_m$ by σ. Find a function f, $\mathbb{R} \times \mathbb{S}^2 \mapsto \mathbb{C}$, which is such that*

(i) $f - \hat{1} \in L^2(\mathbb{R} \times \mathbb{S}^2)$;

(ii) $f(k)$, $\mathbb{R} \mapsto L^2(\mathbb{S}^2)$, is the boundary value of an analytic function meromorphic in $\mathbb{C}^+$ such that $\lim_{|k|\to\infty} \|f(k) - \hat{1}\| = 0$; [Here $\|\cdot\|$ is the norm in $L^2(\mathbb{S}^2)$.]

(iii) In $\mathbb{C}^+$ this analytic function has a finite number of simple poles at the points $i\kappa_m$ and its residue at $i\kappa_m$ lies in the space $\mathscr{H}_m$.

(iv) on $\mathbb{R}$ f satisfies the equation

$$f^* = \Omega f,\tag{2.33}$$

where $f^*(k) = f(-k)$.

Theorem 1.5.7 asserts that if A corresponds to an underlying normal potential that is in the class there assumed and that produces bound states with eigenvalues $-\kappa_m^2$ the sum of whose multiplicities equals n then the spaces $\mathscr{H}_m$ are uniquely determined and can be obtained by the procedure there given. Therefore in that case $A \in \mathscr{A}_n$ and σ is determined by A.

To solve this problem we proceed as in the case without bound states, except that the sought solution to (2.33) now has simple poles in $\mathbb{C}^+$ whose residues lie in known spaces. Thus the Fourier transform (2.9) of $f - \hat{1}$ no longer vanishes for $\alpha < 0$. While we still have (2.10) as the Fourier transform of (2.33), the form of the function η for $\alpha < 0$ is obtained from (1.31) as

$$\eta(\alpha) = \sum_{m,b} p_m^b y_m^b(-\alpha). \tag{2.34}$$

The sum here is over the bound states (whose eigenvalues are $-\kappa_m^2$) and their degeneracies (denumerated by b); the functions $y_m^b(\alpha)$ are explicitly given by

$$y_m^b(\alpha, \theta, x) = Y_{\kappa_m}^b(-\theta)e^{\kappa_m(\theta \cdot x - \alpha)}, \tag{2.35}$$

where the functions Y_κ^b are the characters. The coefficients p_m^b depend on x and are related to the eigenfunctions $u_\kappa^a(x)$ of (1.1) by

$$p_m^b(x) = \frac{1}{2\kappa} \sum_a d_{ba}^{\kappa_m} u_{\kappa_m}^a(x), \tag{2.36}$$

where the matrix d_κ whose entries are d_{ab}^κ is the inverse of the matrix e_κ defined in (1.27).

Equation (2.10) is again separated into the part for $\alpha > 0$,

$$\eta(\alpha) = g(\alpha) + Q\eta(-\alpha) \; + \; \int_0^\infty d\beta \, G(\alpha + \beta)\eta(\beta)$$
$$+ \; \int_0^\infty d\beta \, G(\alpha - \beta)\eta(-\beta),$$

and the part for $\alpha < 0$, in which we change $\alpha \to -\alpha$ so that for $\alpha > 0$

$$\eta(\alpha) = -Qg(-\alpha) + Q\eta(-\alpha) \; - \; \int_0^\infty d\beta \, QG(\beta - \alpha)\eta(\beta)$$
$$- \; \int_0^\infty d\beta \, QG(-\alpha - \beta)\eta(-\beta).$$

The use of (2.34) for $\alpha < 0$ leads to the two equations for $\alpha > 0$:

$$\eta(\alpha) = g(\alpha) \; + \; \sum_{m,b} p_m^b \left[Q y_m^b(\alpha) + \int_0^\infty d\beta \, G(\alpha - \beta) y_m^b(\beta) \right]$$
$$+ \; \int_0^\infty d\beta \, G(\alpha + \beta)\eta(\beta),$$

$$\eta(\alpha) = -Qg(-\alpha) \;+\; \sum_{m,b} p^b_m Q \left[y^b_m(\alpha) - \int_0^\infty d\beta\, G(-\alpha - \beta) y^b_m(\beta) \right]$$
$$- \int_0^\infty d\beta\, QG(\beta - \alpha)\eta(\beta),$$

which we write more simply by means of the operators defined by (2.16) and (2.18): On $\mathbb{R}_+ \times S^2$,

$$\eta = g + [(Q + \mathcal{H})\tilde{y}]p + \mathcal{G}\eta, \tag{2.37}$$

$$\eta = -Qg^* + Q[(1 - \mathcal{G}^*)\tilde{y}]p - Q\mathcal{H}^*\eta, \tag{2.38}$$

where p is the column matrix whose entries are p^b_m, y is the column matrix whose entries are y^b_m, and $\tilde{y}$ is the transpose of y.

Suppose now that $\mathcal{G}^*$ has the eigenvalue -1 with the multiplicity n_σ, where n_σ is the number defined in $\mathcal{H}^1_\sigma$. We shall label the eigenfunctions by r, call them z_r, and denote the column matrix with these entries by z. There are as many z's as y's. Taking the inner product $\langle \cdot, \cdot \rangle_+$ on $L^2(\mathbb{R}_+ \times S^2)$ of (2.38) with Qz_r we obtain

$$sp = c, \tag{2.39}$$

where s is the square matrix and c is the column matrix whose elements are

$$s_{r,m'b'} = \langle z_r, (1 - \mathcal{G}^*)y^{b'}_{m'} \rangle_+ = 2\langle z_r, y^{b'}_{m'} \rangle_+, \tag{2.40}$$

$$c_r = \langle z_r, g^* \rangle_+. \tag{2.41}$$

If s^{-1} exists (see below) we have

$$p = s^{-1}c. \tag{2.42}$$

We then define the projection P and the degenerate operator Λ by

$$\Lambda \xi \;\; := \;\; \tilde{y}s^{-1}\langle z, \xi \rangle_+ \tag{2.43}$$

$$P \;\; := \;\; 1 - (1 - \mathcal{G}^*)\Lambda = P^2. \tag{2.44}$$

Use of (2.42) in (2.37) and (2.38) yields the equations

$$\eta = g + (Q + \mathcal{H})\tilde{y}s^{-1}c + \mathcal{G}\eta, \tag{2.45}$$

$$(Q + \mathcal{H}^*)\eta = -Pg^*. \tag{2.46}$$

If $\mathcal{G}$ does not have the eigenvalue 1 then (2.45) has a unique solution. To prove that this solution also solves (2.46), let us define

$$\xi := (Q + \mathcal{H}^*)\eta + Pg^*,$$

where η solves (2.45). Then, using (2.27), (2.28), and (2.31) we find that $(1 + \mathcal{G}^*)\xi = 0$ as well as $\langle z_r, \xi \rangle_+ = 0$. Therefore, since the functions z_r span the nullspace of $1 + \mathcal{G}^*$, it follows that $\xi = 0$. Thus if η solves (2.45) then it also solves (2.46). Therefore, on the assumption that $\mathcal{G}^*$ has the eigenvalue -1 with the multiplicity n_σ, if $\mathcal{G}$ does not have the eigenvalue 1 and s is invertible, then the system (2.37) and (2.38) for the pair η, p has a unique solution.

Conversely, suppose that the system (2.37) and (2.38) has a unique solution. Then its homogeneous version cannot have a nontrivial solution. Suppose that there is a vector $q \neq 0$ such that $sq = 0$. It then follows that the homogeneous version of (2.37) and (2.38) for the pair η, p has a nontrivial solution. Hence it follows that $sq = 0$ cannot have a nontrivial solution and also $\mathscr{G}$ cannot have the eigenvalue 1. Furthermore, suppose that the multiplicity n of the eigenvalue -1 of $\mathscr{G}^*$ is not equal to n_σ. If $n > n_\sigma$ then the set of linear inhomogeneous algebraic equations (2.39) is larger than the number of unknowns and there is no solution. If $n < n_\sigma$ then the numbers p_m^b are not uniquely determined unless additional linearly independent vectors z_r are used which are not in the nullspace of $\mathbb{1} + \mathscr{G}^*$. In that case the operator $\mathscr{G}$ in (2.45) is replaced by

$$\mathscr{G}^B := \mathscr{G} + (Q + \mathscr{H})\Lambda(Q + \mathscr{H}^*). \tag{2.47}$$

By (2.27) and (2.28) we find that

$$(\mathbb{1} - \mathscr{G}^B)(Q + \mathscr{H}) = (Q + \mathscr{H})P'(\mathbb{1} + \mathscr{G}^*), \tag{2.48}$$

where

$$P' := \mathbb{1} - \Lambda(\mathbb{1} - \mathscr{G}^*) \tag{2.49}$$

is a projection. Since $(\mathbb{1} - P')(\mathbb{1} + \mathscr{G}^*) \neq 0$, it easily follows that $\mathscr{G}^B$ must have the eigenvalue 1 and (2.45) does not have a unique solution. Furthermore, the inhomogeneous term in (2.45) is in $L^2(\mathbb{R}_+ \times S^2)$ since g is in $L^2(\mathbb{R}_+ \times S^2)$ and

$$\begin{aligned}\langle (Q + \mathscr{H})y_m^b, (Q + \mathscr{H})y_m^b\rangle_+ &= \langle y_m^b, (Q + \mathscr{H}^*)(Q + \mathscr{H})y_m^b\rangle_+ \\ &= \langle y_m^b, (\mathbb{1} - \mathscr{G}^{*2})y_m^b\rangle_+ < \infty.\end{aligned}$$

Therefore so is the solution. We therefore have the following result.

Lemma 2.3.9. *The system of equations (2.37) and (2.38) has a unique solution if and only if the operator $\mathscr{G}$ does not have the eigenvalue 1, the operator $\mathscr{G}^*$ has the eigenvalue -1 with an n_σ-dimensional eigenspace, and the matrix s defined by (2.40) is invertible. The solution is given by the solution of the generalized Marchenko equation (2.45) together with (2.42).*

We also note the following.

Lemma 2.3.10. *If the functions in the generalized Marchenko equation (2.45) are as previously defined, the equation always has a solution in $L^2(\mathbb{R}_+ \times S^2)$, even if its homogeneous version has a nontrivial solution.*

Proof. Suppose that σ solves the homogeneous version of (2.45): $\mathscr{G}\sigma = \sigma$. A straightforward computation, using (2.28), (2.29), (2.30), and (2.31), leads to the result $(\mathbb{1} + \mathscr{G})[g + (Q + \mathscr{H})\tilde{y}s^{-1}c] = -(Q + \mathscr{H})Pg^*$. Therefore,

$$2\langle \sigma, [g + (Q + \mathscr{H})\tilde{y}s^{-1}c]\rangle_+ = \langle \sigma, (\mathbb{1} + \mathscr{G})[g + (Q + \mathscr{H})\tilde{y}s^{-1}c]\rangle_+ = 0$$

because by (2.26) $\mathscr{G}\sigma = \sigma$ implies $(Q + \mathscr{H}^*)\sigma = 0$ and $\mathscr{H}^* = \mathscr{H}^\dagger$. Consequently equation (2.45) has a solution even if its homogeneous version has a nontrivial solution. $\square$

Now (2.45) and (2.46), together with the definition (2.42) of p, are equivalent to (2.37) and (2.38), which in turn are equivalent to solving $H_\sigma^1(S_x)$ by means of the inverse Fourier transform of (2.9) with (2.34), i.e.,

$$f(k) = 1 + \int_0^\infty d\alpha\, \eta(\alpha) e^{ik\alpha} + \sum_{m,b} \frac{p_m^b y_m^b(0)}{i(k - i\kappa_m)}. \tag{2.50}$$

If $\mathscr{G}$ has the eigenvalue 1 then, of course, the solution of (2.45) is not unique. Let σ be an eigenfunction of $\mathscr{G}$ with eigenvalue 1, i.e., a solution of the homogeneous version of (2.45). Then it follows from (2.26) that $(Q + \mathscr{H}^*)\sigma = 0$ and hence σ also solves the homogeneous version of (2.46). Thus its right-hand Fourier transform

$$f_0(k) = \int_0^\infty d\alpha\, \sigma(\alpha) e^{ik\alpha} + \sum_{m,b} \frac{p_m^b y_m^b(0)}{i(k - i\kappa_m)} \tag{2.51}$$

solves $H_\sigma^0(S_x)$, which is related to $H_\sigma^1(S_x)$ as $H_0^0(S_x)$ is to $H_0^1(S_x)$. We therefore have the analogue of Lemma 2.3.5.

We collect our results in the following.

Theorem 2.3.11. *The Riemann-Hilbert problem $H_\sigma^1(S_x)$ has a unique solution if and only if $\mathscr{G}$ does not have the eigenvalue 1, the operator $\mathscr{G}^*$ has the eigenvalue -1 with an n_σ-dimensional eigenspace, and the matrix s defined by (2.40) is invertible. This solution is obtained by (2.50) from the solution of the generalized Marchenko equation (2.45) together with (2.42).*

2.4 The x-dependence

Since the input function G in the generalized Marchenko equation depends on $x \in \mathbb{R}^3$, so will the solution. Up to this point this parametric x-dependence has played no role and we have not even found it necessary to indicate it in the notation. However, the function ζ that satisfies the Riemann-Hilbert problem solved in the previous section is supposed to solve the partial differential equation (1.53) so that ψ^+ satisfies (1.1). We must now study the consequences of this for the Fourier transform of ζ, i.e., for the function η.

Taking the Fourier transform of the Lippmann-Schwinger equation (1.3) we obtain by the use of (2.34), (2.35), (2.36), and

$$\hat{\eta}(\alpha, \theta, x) := \eta(\alpha - x \cdot \theta, \theta, x) = \frac{1}{2\pi} \int_{-\infty}^\infty dk\, e^{-ik\alpha} [\psi^+(k, \theta, x) - e^{ikx \cdot \theta}], \tag{2.52}$$

$$\eta(\alpha, \theta, x) = \sigma(\alpha, \theta, x) - \int_\Omega dy\, \frac{V(y)}{4\pi|x - y|} \eta(\alpha - |x - y| + \theta \cdot (x - y), \theta, y)$$
$$- \int_{\Omega'} dy\, \frac{V(y)}{4\pi|x - y|} \sum_{mb} p_m^b(y) y_m^b(-\alpha, \theta, y) e^{\kappa_m[\theta \cdot (x - y) - |x - y|]}, \tag{2.53}$$

where

$$\sigma(\alpha, \theta, x) = -\int_\Omega dy\, \frac{V(y)}{4\pi|x - y|} \delta[\alpha - |x - y| + \theta \cdot (x - y)].$$

The integrals are extended over the interior and exterior (here denoted by Ω'), respectively, of the paraboloid

$$\Omega(\alpha, \theta, x) = \{y \in \mathbb{R}^3 \mid \alpha - |x - y| + \theta \cdot (x - y) > 0\}.$$

As $\alpha \downarrow 0$, the paraboloid Ω shrinks to a point and we have

$$\lim_{\alpha \downarrow 0} \sigma(\alpha, \theta, x) = -\frac{1}{2} \int_0^\infty dt \, V(x - \theta t).$$

Consequently

$$\begin{aligned}
\lim_{\alpha \downarrow 0} \eta(\alpha, \theta, x) &= -\frac{1}{2} \int_0^\infty dt \, V(x - \theta t) \\
&\quad - \int_{\mathbb{R}^3} dy \, \frac{V(y)}{4\pi|x - y|} \sum_{mb} e^{\kappa_m[\theta \cdot (x-y) - |x-y|]} p_m^b(y) y_m^b(0, \theta, y) \\
&= -\frac{1}{2} \int_0^\infty dt \, V(x - \theta t) + \widetilde{y}(0, \theta, x) p(x).
\end{aligned} \tag{2.54}$$

Differentiation of this equation leads to

$$-2\theta \cdot \nabla[\eta(0+, \theta, x) - \widetilde{y}(0, \theta, x)p(x)] = V(x). \tag{2.55}$$

Here y is defined by (2.35). Furthermore, differentiation of equation (2.53) leads to the hyperbolic partial differential equation for $\alpha > 0$

$$\left[\Delta - 2\frac{\partial}{\partial \alpha}\theta \cdot \nabla - V(x)\right]\eta(\alpha, \theta, x) = 0 \tag{2.56}$$

and equation (2.34) for $\alpha < 0$, or

$$\left[\Delta - \frac{\partial^2}{\partial \alpha^2} - V(x)\right][\widehat{\eta}(\alpha, \theta, x) - \widetilde{y}(\alpha - \theta \cdot x, \theta, x)p(x)] = 0 \tag{2.57}$$

for $\alpha > \theta \cdot x$, and

$$\widehat{\eta}(\alpha, \theta, x) = \sum_b p_m^b(x) Y_{\kappa_m}^b(-\theta)e^{\kappa_m \alpha} \tag{2.58}$$

for $\alpha < \theta \cdot x$, with the boundary condition obtained from (2.55)

$$-2\theta \cdot \nabla[\widehat{\eta}(\alpha = \theta \cdot x+, \theta, x) - \widetilde{y}(0, \theta, x)p(x)] = V(x). \tag{2.59}$$

These equations can alternatively be obtained by the use of (2.52) together with (1.1). Note that the function

$$y_m^b(\alpha - \theta \cdot x, \theta, x) = Y_{\kappa_m}^b(-\theta)e^{\kappa_m \alpha}$$

is such that the second term in the second bracket of (2.57) may be deleted.

The following result, needed later in the proof of Lemma 2.4.6, will be proved in Section 2.4.1.

Lemma 2.4.1. *The system (2.57) and (2.59) has at most one solution.*

From the point of view of the direct scattering problem the equations (2.57) and (2.59) may be regarded as equivalent to the Schrödinger equation (1.1), and (2.53) as the equivalent of the Lippmann-Schwinger equation(1.3), although the precise sense in which they hold has not been established. From the point of view of the inverse problem equation (2.55) gives a simple formula for the potential in terms of η, which is computed by means of the generalized Marchenko equation, provided that such an underlying potential exists. Alternatively, of course, (2.56) may be used to extract the potential. The function

$$
\begin{aligned}
\rho(k, \theta, x) \;=\; & e^{ik\theta \cdot x} + \int_{x \cdot \theta}^{\infty} d\alpha \, \widehat{\eta}(\alpha, \theta, x) e^{ik\alpha} \\
& + \sum_{m,b} \frac{p_m^b(x) Y_{\kappa_m}^b(-\theta)}{i(k - i\kappa_m)} e^{i(k - i\kappa_m)\theta \cdot x}
\end{aligned}
\tag{2.60}
$$

satisfies the Schrödinger equation with that potential. Looked at in this way, (2.55) has a remarkable characteristic: the function η obtained as a solution of the generalized Marchenko equation must be such that the left-hand side of (2.55) is independent of the direction θ, because the right-hand side is so. For a solution of (2.13) or (2.45) to have this property appears as a miracle, and it will be called by that name. It is the constraint on the data needed to ensure that the five-variable function $A(k, \theta, \theta')$ is in one-to-one correspondence with an underlying function $V(x)$ of three variables. Unfortunately no method is known that would lead to a direct recognition of an admissible function A, that is, one that is, in fact, the scattering amplitude of some potential.

Suppose now that a function A is given that is not known to be admissible, and the generalized Marchenko equation (2.13) or (2.45) is solved. Let $\eta(\alpha, \theta, x) = \widehat{\eta}(\alpha + \theta \cdot x, \theta, x)$ be a solution and assume that the miracle occurs, i.e., the left-hand side of (2.55) is independent of θ. It is incumbent upon us to check whether η, as function of x, satisfies (2.56), where V is the function defined by (2.55). In order to do that we apply the operator $(\Delta - 2\frac{\partial}{\partial \alpha}\theta \cdot \nabla)$ to (2.37) and (2.38). The ensuing computation is facilitated by the formula

$$
\left(\Delta - 2\frac{\partial}{\partial \alpha}\theta \cdot \nabla \right) [\mathsf{G}(\alpha, \theta, \theta')f(x)] = \left[\mathsf{G}(\alpha, \theta, \theta')\Delta + 2\frac{\partial}{\partial \alpha}\mathsf{G}(\alpha, \theta, \theta')\theta' \cdot \nabla \right] f(x)
$$

for any function $f(x)$, which is readily verified by means of (2.8) and (2.5). As a result one finds, after integrating by parts, that if $V(x)$ is defined by

$$
V(x) = -2\theta \cdot \nabla[\eta(0+, \theta, x) - \widehat{y}(0, \theta, x)p(x)]
$$

where y is defined by (2.35), and if the right-hand side of this formula is independent of θ, then the function

$$
\xi(\alpha, \theta, x) := \left[\Delta - 2\frac{\partial}{\partial \alpha}\theta \cdot \nabla - V(x) \right] \eta(\alpha, \theta, x)
$$

satisfies the equations for $\alpha > 0$

$$
\begin{aligned}
\xi &= [(Q + \mathscr{H})\widehat{y}]w + \mathscr{G}\xi, \\
\xi &= [Q(1 - \mathscr{G}^*)\widehat{y}]w - Q\mathscr{H}^*\xi,
\end{aligned}
$$

where $w_m^b := [\varDelta - \kappa_m^2 - V(x)]p_m^b(x)$. These equations are exactly the homogeneous versions of (2.37) and (2.38); they lead directly to the homogeneous form of (2.45) and (2.46). Therefore, if the system (2.37), (2.38) has a unique solution, then $\xi = 0$; hence η satisfies the hyperbolic equation (2.56) and $w = 0$, which means that p_m^b satisfies the Schrödinger equation with the potential $V(x)$ and the eigenvalue $-\kappa_m^2$, though we do not know yet whether $p_m^b(x)$ is in $L^2(\mathbb{R}^3)$. (See Lemma 2.4.6 for that.)

We therefore have the following result.

Theorem 2.4.2. *If for the given scattering amplitude there exists a normal potential (see Definition 1.2.5) in $\mathscr{W}$ then the generalized Marchenko equation (2.45) has a solution that is miraculous, i.e., for which the left-hand side of (2.55) is independent of θ. Conversely, if $\mathscr{G}$ does not have 1 as an eigenvalue $\mathbb{1}+\mathscr{G}^*$ has an n_σ-dimensional nullspace, s of (2.40) is invertible, and the solution of the generalized Marchenko equation (2.45) or (2.13) is miraculous, then the right-hand Fourier transform as in (2.60) or (2.25) satisfies the Schrödinger equation with the potential (2.55) and the functions $p_m^b(x)$ given by (2.42) satisfy (1.1) with the same potential and the eigenvalues $-\kappa_m^2$.*

Remark 1. There is no known assurance that, if an underlying potential exists, the corresponding operator $\mathscr{G}$ cannot have the eigenvalue 1. In the one-dimensional case there is such assurance for an analogous question, but for higher dimensions this is one of the important open questions. At the present time it appears to be possible that a potential exists for a given scattering amplitude but because the homogeneous form of the generalized Marchenko equation has a nontrivial solution, this potential cannot be uniquely recognized by a simple examination of the solution of the generalized Marchenko equation. [Of course, it could be recognized by testing all the solutions to see whether they satisfy (1.1).]

Remark 2. It is understood in Theorem 2.4.2 that the bound-state data (their number, degeneracies, and character spaces) that enter into (2.45) are constructed from the scattering amplitude in the canonical manner explained earlier. The connection of these data to the forward scattering amplitude is a necessary condition for a normal potential in $\mathscr{W}$ to exist.

Is it possible to ensure by a condition on the scattering amplitude A that $\|\mathscr{G}\| < 1$ rather than just $\|\mathscr{G}\| \le 1$, which is always the case? We do have the following.

Lemma 2.4.3. *If for each $k \in \mathbb{R}$, $\|kA(k)\|_{\mathbb{S}^2} < c$, where $\|\cdot\|_{\mathbb{S}^2}$ is the operator norm on $L^2(\mathbb{S}^2)$, then for all $x \in \mathbb{R}^3$, $\|\mathscr{G}\|_{\mathbb{S}^2} < c/2\pi$, where $\|\cdot\|$ is the operator norm on $L^2(\mathbb{R}_+ \times \mathbb{S}^2)$.*

Proof. Take $\xi \in L^2(\mathbb{R}_+ \times \mathbb{S}^2)$, $\xi(\beta) = 0$ for $\beta < 0$ and $f(k) := \int_0^\infty e^{ik\beta} \xi(\beta)$,

$$v(\alpha) := \int_0^\infty d\beta\, \mathsf{G}(\alpha + \beta)\xi(\beta),$$

where G is given by (2.24). Then

$$v(\alpha) = \frac{1}{2\pi} \int_{-\infty}^\infty dk\, \omega(k) e^{-ik\alpha},$$

where $\omega(k) = (ik/2\pi)QA_x(k)f(-k)$, and

$$\|v\|_+^2 := \int_0^\infty d\alpha \, \|v(\alpha)\|_{S^2}^2 \leq \int_{-\infty}^\infty d\alpha \, \|v\|_{S^2}^2 = \frac{1}{2\pi} \int_{-\infty}^\infty dk \, \|\omega\|_{S^2}^2,$$

which means

$$\|\mathcal{G}\xi\|_+^2 \leq \frac{1}{(2\pi)^3} \int_{-\infty}^\infty dk \, k^2 \|A_x(k)f(-k)\|_{S^2}^2 \leq \frac{c^2}{(2\pi)^3} \int_{-\infty}^\infty dk \, \|f\|_{S^2}^2$$

$$= \frac{c^2}{(2\pi)^2} \|\xi\|_+^2$$

because $\|A_x\| = \|A\|$. $\square$

This lemma implies that if A is uniformly small then $\mathcal{G}$ is small for all x. Furthermore,

$$\|A\|_{S^2}^2 \leq \int_{S^2 \times S^2} d\theta d\theta' \, |A(k,\theta,\theta')|^2 = 4\pi\bar{\sigma},$$

where $\bar{\sigma}$ is the average total scattering cross section. Therefore, if for all k we have $k^2\bar{\sigma} < \pi$ then it follows that $\|\mathcal{G}\| < 1$. Note also that if $\|kA(k)\|_{S^2} < 2\pi$ (which ensures $\|\mathcal{G}\| < 1$), then each eigenphase shift is less than π.

Equation (2.54) may be written in the form

$$\eta(0-,\theta,x) - \eta(0+,\theta,x) = \tfrac{1}{2}v(x,\theta),$$

where

$$v(x,\theta) := \int_0^\infty dt \, V(x - \theta t),$$

so that $V(x) = \theta \cdot \nabla v(x,\theta)$. Going back to equation (2.10) and letting $\alpha \to 0$ from above and from below one obtains

$$\eta(0\pm) = Q\eta(0\mp) + g(0\pm) + \int_{-\infty}^\infty d\alpha \, G(\alpha)\eta(\alpha), \tag{2.61}$$

which implies

$$\begin{aligned} g(0-) - g(0+) &= (\mathbb{1} + Q)[\eta(0-) - \eta(0+)] \\ &= \tfrac{1}{2}(\mathbb{1} + Q)v = \frac{1}{2} \int_{-\infty}^\infty dt \, V(x - \theta t). \end{aligned} \tag{2.62}$$

Thus $g(\alpha)$ is discontinuous at $\alpha = 0$, and its discontinuity is directly related to $(\mathbb{1}+Q)v$. This discontinuity comes entirely from the large-k limit of the scattering amplitude, since no finite piece of the integral in (2.8) contributes to it. Equation (2.62) is a direct consequence of the fact that the scattering amplitude approaches the Born approximation for large k. In fact, (2.62) is easily verified directly if A in (2.62) is replaced by its Born approximation. Since V can be recovered from a knowledge of the right-hand side of (2.62) [see DR85] the discontinuity of $g(\alpha)$ at $\alpha = 0$ is sufficient to recover V. However, this solution of the inverse problem is equivalent to the solution via the large-k limit of A; since it is invariant under changes in A at any finite k it cannot lead to a characterization of admissible functions A.

Applying the operation $\theta \cdot \nabla$ to the two equations (2.61) and subtracting the results gives

$$2\theta \cdot \nabla[g(0+) - g(0-)] = (Q - \mathbb{1})\theta \cdot \nabla v.$$

By (2.62) and the miracle the right-hand side of this equation vanishes. Therefore, if a potential exists then, although $g(\alpha)$ has a discontinuity at $\alpha = 0$, $\theta \cdot \nabla g(\alpha)$ is continuous there.

One may also take the average value of v,

$$\tilde{v}(x) := \int_{S^2} d\theta\, v(\theta) = \int_{S^2} d\theta \int_0^\infty dt\, V(x - \theta t) = \int_{\mathbb{R}^3} dy\, \frac{V(y)}{|x - y|^2}. \qquad (2.63)$$

A knowledge of this function is sufficient to construct V uniquely, either by Fourier transformation or by the method of Saitō. (See Section 2.9.)

Theorem 2.4.4. *The functions A and $\overline{A}$ cannot both be admissible scattering amplitudes unless $A = 0$.*

Proof. Let A_1 and A_2 be two scattering amplitudes in $\mathscr{A}$ such that $A_2 = \overline{A}_1$. This implies that the two functions G_1 and G_2 that correspond to them by (2.8) are such that $G_2(\alpha) = -G_1(-\alpha)$. By (2.62) and (2.63) therefore $\tilde{v}_1 = \tilde{v}_2$, and since $\tilde{v}$ uniquely determines V, it follows that $V_1 = V_2$ if both potentials exist. But then $A_1 = A_2$ and therefore $A_1 = \overline{A}_1$, which by the unitarity of the S matrix implies that $A = 0$. $\square$

We have now constructed a potential and a solution of the Schrödinger equation on the basis of a given function $A(k, \theta, \theta')$. However, there is no assurance yet that the scattering amplitude that corresponds to this potential actually is the function A that formed the starting point. Whereas a potential leads to a unique scattering amplitude via the solution of the direct scattering problem it cannot be ruled out a priori that two different functions G, obtained from two different functions A, lead to the same potential via the solution of the generalized Marchenko equation.

One answer to this question in an immediate consequence of Lemma 1.5.1, as was (1.61) [RW87, see also Ra87c]:

Theorem 2.4.5. (Ramm-Weaver) *The following two conditions are equivalent:*
(A) The scattering amplitude A corresponds to a potential in $\mathscr{W}$ that causes no bound states;
(B) The generalized Marchenko equation (2.13) has a solution whose right-handed Fourier transform as in (2.25) satisfies the Schrödinger equation (1.1) and has the asymptotic form (1.2).
Furthermore, if (A) holds then there is exactly one solution of (2.13) that satisfies (B).

(Actually, the weaker condition in (A) that $|V(x)| + |\nabla V(x)| \leq c(1 + |x|)^{-a}$, $a > 3$, stated in the original version of this theorem in [RW87], suffices.) In other words, this theorem says that if the function obtained by (2.13) and (2.25) solves (1.1) and has the asymptotic form (1.2), then the scattering amplitude in (1.2) coincides with the function A in (2.13). It closes the circle that leads from A to V and back to A and may serve to characterize admissible functions A. It is, however, of limited utility, because if one ascertains that all the conditions in (B) are satisfied, one may as well check one further condition, namely, whether the function A in (1.2) agrees with the input function A in (2.13). So formulated, the theorem would, of course, be a tautology.

Our aim, instead, is to find conditions to ensure the closing of the circle $A \longrightarrow V \longrightarrow A$ that refer directly to the input function in the generalized Marchenko equation. We have the following result, which will be proved in Section 2.4.1.

Lemma 2.4.6. *If $A \in \mathcal{A}$ and the unique solution of the generalized Marchenko equation (2.45) leads, via (2.60), to a solution of the Schrödinger equation then this solution satisfies (1.2) and the scattering amplitude A in (1.2) coincides with the input function in (2.45). Furthermore, the functions $p_m^b(x)$ given by (2.42) are L^2-eigenfunctions of the Schrödinger equation and the functions $Y_{\kappa_m}^b$ are the characters.*

The circle from the potential to the scattering amplitude and back to the potential, starting either at the potential or at the scattering amplitude has now been closed. If $\mathcal{G}$ does not have the eigenvalue 1, $\mathcal{G}^*$ has the eigenvalue -1 with the multiplicity n_σ, the matrix s of (2.40) is invertible and the solution of (2.45) is miraculous, then the right-hand Fourier transform (2.60) of this solution satisfies the Schrödinger equation and the outgoing-wave boundary condition (1.2), and the corresponding scattering amplitude coincides with the input function A.

Theorem 2.4.7. (Sufficient Condition) *If (i) $A \in \mathcal{A}$, (ii) 1 is not in the spectrum of $\mathcal{G}$, but it is in the spectrum of $-\mathcal{G}^*$ with the multiplicity n_σ, the matrix s of (2.40) is invertible and (iii) the solution of the generalized Marchenko equation (2.45) is miraculous, i.e., the left-hand side of (2.55) is independent of θ, then the right-hand Fourier transform (2.60) satisfies the Schrödinger equation with the potential given by (2.55). Furthermore, the scattering amplitude corresponding to this potential equals the input function A, the functions p_m^b are eigenfunctions corresponding to the eigenvalues $-\kappa_m^2$, and the functions $Y_{\kappa_m}^b$ are the characters.*

The reason why this is not a full characterization is that (a) it is not known whether the map from $A \in \mathcal{A}$ necessarily leads to $V \in \mathcal{W}$, and (b) it is not known whether the condition that 1 not be in the spectrum of $\mathcal{G}$ is necessary for all $V \in \mathcal{W}$.

2.4.1 Proofs

Proof of Lemma 2.4.1. The difference f between two solutions must satisfy the system, for $\alpha \geq z$

$$\left(\Delta - \frac{\partial^2}{\partial \alpha^2} - V \right) f = 0,$$

$$\left(\frac{\partial}{\partial z} + \frac{\partial}{\partial \alpha} \right) f \bigg|_{\alpha = z} = 0,$$

if we choose the z-axis in the direction of θ. We introduce characteristic coordinates $\xi = \frac{1}{2}(z + \alpha)$, $\zeta = \frac{1}{2}(z - \alpha)$, so that the system reads

$$\frac{\partial^2}{\partial\xi\partial\zeta} f = (V - \Delta_2)f,$$

$$\frac{\partial}{\partial\xi} f \Big|_{\zeta=0} = 0,$$

where Δ_2 is the Laplacian in x and y. Introduce bounded solutions in $\mathbb{R}^2$ of

$$(V - \Delta_2)g = k^2(z)g$$

for k^2 in the corresponding continuous spectrum, at fixed z. We expand the function f on this basis and obtain the system

$$\frac{\partial^2(cg)}{\partial\xi\partial\zeta} = k^2cg,$$

$$\frac{\partial(cg)}{\partial\xi} \Big|_{\zeta=0} = 0,$$

where k and g are given and depend on $z = \xi + \zeta$. After one sets $h = \partial cg/\partial\xi$ the system reads

$$\frac{\partial h}{\partial\zeta} = k^2 \int_{-\infty}^{\xi} d\xi' h, \qquad h \Big|_{\zeta=0} = 0,$$

which has only the trivial solution $h = 0$. Therefore the function $cg = c(\xi, \zeta)$ $g(x, y, \xi + \zeta)$ is independent of ξ, which implies that either $c = 0$ or the eigenfunctions g are independent of z. But that is possible only if, for that particular direction of θ, the potential V is independent of z. Since V vanishes at infinity in all directions, that cannot be the case. Therefore $c = 0$ and hence $f = 0$. □

For Lemma 2.4.6 we shall first give a direct argument that lacks sufficient rigor to be quite convincing. It will, however, show the role played by forward analyticity. It will be followed by a more indirect proof.

Let us begin by assuming that there are no bound states and deal with equation (2.23), for $\alpha > 0$,

$$\eta(\alpha - \theta \cdot \dot{x}, \theta, x) = \int d\theta'\, G(\alpha + \theta' \cdot x, \theta, \theta')$$

$$+ \int_{S^2} d\theta' \int_0^\infty d\beta\, G(\alpha + \beta + \theta' \cdot x, \theta, \theta')\eta(\beta, \theta', x), \quad (2.64)$$

where G is given by (2.24). We wish to prove that for all $\alpha \in \mathbb{R}$, $x \in \mathbb{R}^3$

$$\lim_{r\to\infty} r\eta(\alpha + r - \theta \cdot x, \theta, x) = \widehat{A}(\alpha, \hat{x}, \theta), \quad (2.65)$$

where $r = |x|$, $\hat{x} = x/|x|$, and

$$\widehat{A}(\alpha, \hat{x}, \theta) = \frac{1}{2\pi} \int_{-\infty}^{\infty} dk\, A(k, \hat{x}, \theta)e^{-ik\alpha}.$$

One readily finds that

$$\lim_{r \to \infty} r \int_{S^2} d\theta' \, G(\alpha + r + \theta' \cdot x, \theta, \theta') f(\theta') = \widehat{A}(\alpha, -\theta, -\hat{x}) f(-\hat{x}),$$

$$\lim_{r \to \infty} r \int_{S^2} d\theta' \, G(\alpha - r + \theta' \cdot x, \theta, \theta') f(\theta') = -\widehat{A}(\alpha, \theta, \hat{x}) f(\hat{x}).$$

Using these limits in (2.64), one obtains for $\theta = \hat{x}$ or $\theta = -\hat{x}$ and $\alpha > 0$

$$\lim_{r \to \infty} r\eta(\alpha, \hat{x}, x) = \widehat{A}(\alpha, -\hat{x}, -\hat{x})$$
$$+ \lim_{r \to \infty} \int_0^\infty d\beta \, \widehat{A}(\alpha + \beta, -\hat{x}, -\hat{x}) \eta(\beta, -\hat{x}, x),$$

$$\lim_{r \to \infty} r\eta(\alpha, -\hat{x}, x) = -\widehat{A}(\alpha, \hat{x}, \hat{x}) - \lim_{r \to \infty} \int_0^\infty d\beta \, \widehat{A}(\alpha + \beta, \hat{x}, \hat{x}) \eta(\beta, \hat{x}, x),$$

and therefore, using the reciprocity property of A,

$$\lim_{r \to \infty} r\eta(\alpha, \hat{x}, x) = \widehat{A}(\alpha, \hat{x}, \hat{x})$$

for $\alpha > 0$. However, if A has forward analyticity then both sides of this equation vanish for $\alpha < 0$, and thus it holds for all $\alpha \neq 0$. We now take $\theta \neq \hat{x}$. Then for all $\alpha \in \mathbb{R}$

$$\lim_{r \to \infty} r\eta(\alpha + r - \theta \cdot x, \theta, x) = \widehat{A}(\alpha, -\theta, -\hat{x})$$
$$+ \lim_{r \to \infty} \int_0^\infty d\beta \, \widehat{A}(\alpha + \beta, -\theta, -\hat{x}) \eta(\beta, -\hat{x}, x) = \widehat{A}(\alpha, \hat{x}, \theta)$$

by reciprocity. Thus we have (2.65) for all $\alpha \in \mathbb{R}$, $\theta, \hat{x} \in S^2$, except possibly for $\alpha = 0$.

We next form the function

$$\psi(k, \theta, x) := e^{ik\theta \cdot x} + \int_{-\infty}^\infty d\alpha \, e^{ik(\alpha + \theta \cdot x)} \eta(\alpha, \theta, x),$$

where η vanishes for $\alpha < 0$. Then

$$\lim_{r \to \infty} re^{-ikr} \left[\psi(k, \theta, x) - e^{ik\theta \cdot x} \right] = \lim_{r \to \infty} \int_{-\infty}^\infty d\alpha \, e^{ik\alpha} r\eta(\alpha + r - \theta \cdot x, \theta, x)$$
$$= \int_{-\infty}^\infty d\alpha \, e^{ik\alpha} \widehat{A}(\alpha, \hat{x}, \theta) = A(k, \hat{x}, \theta),$$

and thus we have (1.2) with the same function A that formed the input in (2.13).

If there are bound states (2.45) must be used instead of (2.13). Since y_m^b and the elements of the matrix s increase like $e^{\kappa_m r}$ as $r \to \infty$ we shall define new matrices s', and y' whose elements are

$$s'_{n,mb} := s_{n,mb} e^{-r\kappa_m},$$
$$y'^b_m := y^b_m e^{-r\kappa_m}.$$

Then $\tilde{y} s^{-1} c = \tilde{y}' s'^{-1} c$, the normalization of the functions z_n drops out, and one readily finds that as $r \to \infty$ in the computation of the function $\eta(\alpha + r - \theta \cdot x, \theta, x)$ from (2.45) the bound-state terms on the right-hand side decrease exponentially. Thus for large r (2.45) becomes equal to (2.13), and the result for the scattering amplitude is the same as before.

Equation (2.42) becomes

$$p_m^b = e^{-r\kappa_m}(s'^{-1}c)_{mb}.$$

Therefore, $p_m^b = O(e^{-\kappa_m r})$ for large r, as is appropriate for an eigenfunction of the Schrödinger equation with eigenvalue $-\kappa_m^2$. Finally, since the scattering amplitude that corresponds to the constructed potential V is the same as the input function A, the residues at the poles of its forward-value, and their angle-derivatives, are also equal and therefore, so are the character spaces.   □

Proof of Lemma 2.4.6. Suppose we are given a scattering amplitude A and thereby also the bound-state eigenvalues, characters, and constants d_κ. We construct the operator $\mathcal{G}$ and solve the generalized Marchenko equation (2.45). Let the unique solution be η. We then construct the functions p_m^a by means of (2.42) and suppose that the miracle happens, i.e., that (2.59) is independent of θ. We then have a potential V, and by Theorem 2.4.2 the functions ψ and p_κ^a obtained from η solve the Schrödinger equation with V. We then solve the direct scattering problem with V and calculate the scattering solution ψ' and the scattering amplitude A'; Fourier transformation of ψ' leads to η'. The question is whether the primed quantities equal the unprimed ones.

Since both η and η' solve the system (2.57), (2.59), according to Lemma 2.4.1 they can differ at most by functions of the form (2.58) taken to $\alpha > \theta \cdot x$: $\widehat{\eta}' - \widehat{\eta} = \widehat{\eta}'' - \widehat{\eta}'''$, where

$$\widehat{\eta}''(\alpha, \theta, x) = \sum_a p_m^a(x) Y_{\kappa_m}^a(-\theta) e^{\kappa_m \alpha},$$

$$\widehat{\eta}'''(\alpha, \theta, x) = \sum_a p_m'^a(x) Y_{\kappa_m}'^a(-\theta) e^{\kappa_m \alpha}.$$

Since neither $\widehat{\eta}$ nor $\widehat{\eta}'$ is exponentially increasing as a function of α, it follows that for each κ_m

$$\sum_a p_m^a(x) Y_{\kappa_m}^a(\theta) = \sum_a p_m'^a(x) Y_{\kappa_m}'^a(\theta).$$

This implies that the functions $p_m^a(x)$ must be in $L^2(\mathbb{R}^3)$ as the functions $p_m'^a(x)$ are, and that the spaces spanned by the functions $Y_{\kappa_m}^a$ and $Y_{\kappa_m}'^a$ are equal. Furthermore, $\widehat{\eta}' = \widehat{\eta}$. Consequently $\psi' = \psi$ and hence $A' = A$.   □

2.5 Variational Principles

In this section the Riemann-Hilbert problem will be formulated by means of variational principles. The following functional will serve that purpose:

$$\widehat{F}(f) := \mathcal{R} - \frac{1}{2\pi} \int_{-\infty}^{\infty} dk \, \|\varphi(k) - Q S_x(k) \varphi(-k)\|^2, \tag{2.66}$$

in which $\varphi(k) = \widehat{1} + f(k)$, f is an L^2-function: $\mathbb{R} \mapsto L^2(S^2)$, $\|\cdot\|$ is the norm on $L^2(S^2)$, and

$$\mathcal{R} := \frac{1}{\pi} \lim_{\alpha \downarrow 0} \int_{-\infty}^{\infty} dk \, (\widehat{1}, (\mathbb{1} - S_x(k)) \widehat{1}) \cos k\alpha = \frac{1}{4\pi^3} \int_0^{\infty} dk \, k^2 \|A_x(k) \widehat{1}\|^2. \tag{2.67}$$

Here $(\cdot, \cdot)$ is the inner product on $L^2(\mathbf{S}^2)$ and A_x is the scattering amplitude of a shifted potential, defined in (1.81). In obtaining the second form of $\mathscr{R}$ from the first, unitarity and reciprocity have been used. By Lemmas 1.5.11 and 1.5.13 $\widehat{F}$ and $\mathscr{R}$ exist for all $f \in L^2(\mathbb{R} \times \mathbf{S}^2)$ if the potential $V \in \mathscr{W}$, and $\widehat{F}(f) \le \mathscr{R}$, with the equality sign holding if and only if φ satisfies equation (2.6) in the Riemann-Hilbert problem. Because $S_x(-k) = \overline{S_x(k)}$, the functions may be restricted to those with the same symmetry, $f(-k) = \overline{f(k)}$.

Let us begin by assuming that there are no bound states. We re-express the functional $\widehat{F}$ as

$$\widehat{F}(f) = 2\widehat{a}(f) - \widehat{b}(f) \tag{2.68}$$

where

$$\widehat{a}(f) \;\; = \;\; \frac{i}{2\pi^2} \int_{-\infty}^{\infty} dk\, k\big(\hat{1}, A_x(k)f(-k)\big), \tag{2.69}$$

$$\widehat{b}(f) \;\; = \;\; \frac{1}{2\pi} \int_{-\infty}^{\infty} dk\, \|f(-k) - QS_x(-k)f(k)\|^2. \tag{2.70}$$

It may also be expressed in terms of the Fourier transform of $f(k)$,

$$\xi(\alpha) \;:=\; \frac{1}{2\pi} \int_{-\infty}^{\infty} dk\, f(-k) e^{ik\alpha}.$$

If the functions $f(k)$ are restricted to be boundary values of analytic functions that are holomorphic in $\mathbb{C}^+$ and vanish there as $|k| \to \infty$ then the Fourier transforms $\xi(\alpha)$ vanish for $\alpha < 0$. In that case

$$\widehat{F}(f) = F^+(\xi) = 2a^+(\xi) - b^+(\xi), \tag{2.71}$$

where

$$a^+(\xi) \;\; := \;\; \widehat{a}(f) = 2\langle \mathsf{g}, \xi \rangle_+, \tag{2.72}$$

$$b^+(\xi) \;\; := \;\; \widehat{b}(f) = 2(\langle \xi, \xi \rangle_+ - \langle \xi, \mathscr{G}\xi \rangle_+), \tag{2.73}$$

$\mathscr{G}$ is the operator defined by (2.16) on $\mathbb{R}_+ \times \mathbf{S}^2$, g is defined by (2.11), and $\langle \cdot, \cdot \rangle_+$ is the inner product on $L^2(\mathbb{R}_+ \times \mathbf{S}^2)$. Each of the functionals a and b is homogeneous, a linear and b bilinear. As f may be restricted to have the symmetry $f(-k) = \overline{f(k)}$ for real k, ξ may be taken to be real.

Suppose that ζ solves the Riemann-Hilbert problem H_0^1. Then the Fourier transform η of $\zeta - \hat{1}$ as in (2.9) satisfies the generalized Marchenko equation (2.20), and as a result $a^+(\eta) = b^+(\eta)$ and $F^+(\eta) = a^+(\eta) \ge 0$. If, near $\xi = \eta$, ξ is varied then to first order in the variation $\delta\xi$,

$$\delta F^+ = 2\langle (\mathsf{g} - \eta + \mathscr{G}\eta), \delta\xi \rangle_+ + 2\langle \delta\xi, (\mathsf{g} - \eta + \mathscr{G}\eta) \rangle_+ = 0.$$

Conversely, since both terms in it are real, this expression shows that if $F^+(\xi)$ is stationary at $\xi = \eta$ for all small variations $\delta\xi$ then η solves the generalized Marchenko equation (2.20). Since $\|\mathscr{G}\| \le 1$, the second variation of F^+ is non-positive:

$$\delta^2 F^+ = -4\langle \delta\xi, (\mathbb{1} - \mathscr{G})\delta\xi \rangle_+ = -2b^+(\delta\xi) \le 0.$$

Thus the point $\xi = \eta$ is a local *maximum* of F^+. If $\mathscr{G}$ does not have the eigenvalue 1 then the extremal point of F^+ is unique and its value there is $a^+(\eta)$. If $\mathscr{G}$ has the eigenvalue 1 then Lemma 2.3.3 says that $a^+(\eta_0) = 0$ for all solutions η_0 of the homogeneous version of (2.20) and F^+ takes on its extremal value for all solutions of (2.20). The functional F^+ restricted to the orthogonal complement of the eigenspace of $\mathscr{G}$ at 1 has its extremum at a unique point.

The local maximum of F^+ is, in fact, a *global* maximum: For all $\xi \in L^2(\mathbb{R} \times S^2)$, $F^+(\xi) \leq F^+(\eta)$, where η solves the generalized Marchenko equation (2.20). That this is the case follows directly from the identity

$$F^+(\xi) = a^+(\eta) - b^+(\xi - \eta), \tag{2.74}$$

which holds for any solution η of (2.20), and which is easily derived from (2.71), (2.72), and (2.73). Since $b^+ \geq 0$, it follows that for all ξ, $F^+(\xi) \leq F^+(\eta) = a^+(\eta)$. That F^+ always has a global maximum follows from Lemma 2.3.3.

As was discussed in detail in Section 2.3, since the equation (2.13) holds for $\alpha > 0$ only, it is not equivalent to the statement that its Fourier transform as in (2.15), satisfies (2.6) and thus solves the Riemann-Hilbert problem. Lemma 2.3.4 gives a sufficient condition for this to be the case. We now have a new necessary and sufficient condition for a solution η of (2.20) to lead to a solution of the Riemann-Hilbert problem by (2.15): φ, defined in terms of η as in (2.15), solves the Riemann-Hilbert problem H_0^1 if and only if $a^+(\eta) = \mathscr{R}$. Since $a^+(\eta_1) = a^+(\eta_2)$ for any two solutions η_1 and η_2 of (2.20), all solutions of (2.20) lead to solutions of H_0^1 if one does. Thus we have the following.

Theorem 2.5.1. *The functional $F^+(\xi)$, which is identical to the functional $\widehat{F}(f)$ restricted to functions that are boundary values of analytic functions holomorphic in $\mathbb{C}^+$ and tending to 1 there at ∞, and where $\varphi = \hat{1} + f$ is related to η by Fourier transformation as in (2.9), always has a global maximum. This maximum occurs at $\xi = \eta$, where η solves the generalized Marchenko equation (2.20), and its value is $F^+(\eta) = a^+(\eta) \geq 0$. At no other point is F^+ stationary. If and only if $a^+(\eta) = \mathscr{R}$, as given by (2.67), φ solves the Riemann-Hilbert problem H_0^1. If (2.20) has more than one solution then F^+ has the same value for all of them, and if, via (2.15), one solves H_0^1 then they all do.*

Another functional is obtained by maximizing $F^+(v\xi)$ with respect to the real constant v. For fixed ξ the extremum occurs at $v = a^+(\xi)/b^+(\xi)$ and the maximum value of $F^+(v\xi)$ is given by

$$\mathscr{F}^+(\xi) = [a^+(\xi)]^2/b^+(\xi). \tag{2.75}$$

This functional is positive and independent of the scale of ξ, as $\mathscr{F}^+(v\xi) = \mathscr{F}^+(\xi)$ for all $v \in \mathbb{R}$. The functional $\mathscr{F}^+(\xi)$ too has a maximum at $\xi = \eta$ if η solves (2.20) and its value there is $a^+(\eta)$. However, $\mathscr{F}^+(\xi)$ takes on its extremal value not only at $\xi = \eta$ but at $\xi = v\eta$ for all $v \in \mathbb{R}$. That the extremum of $\mathscr{F}^+(\xi)$ at $\xi = \eta$ is a global maximum follows immediately from the fact that by (2.74) $\mathscr{F}^+(\xi) = f^+(v\xi) = \mathscr{F}^+(\eta) - b^+(v\xi - \eta)$ with $v = a^+(\xi)/b^+(\xi)$. Thus η can be computed by maximizing $\mathscr{F}^+(\xi)$, in which the normalization of ξ may be held fixed. However, having found that $\mathscr{F}^+$ is maximal at $\xi = \sigma$ we can conclude

point of zeros of $F(k)$. Even though $F(k)$ is analytic at k_0, there may exist a sequence of points $k_n \in \mathbb{C}^+$, accumulating at k_0, and a corresponding sequence of linearly independent vectors a_n, such that $F(k_n)a_n = 0$ for each n. In such a case $F^{-1}(k)$ will have an essential singularity at k_0.

Suppose that problem $W_0^1(S)$ has a solution F and that $W_0^1(S^*)$ has a solution F'. Then it follows from the properties of unitarity, reciprocity, and k-symmetry [i.e., (ii), (iii), and (iv) of Definition 1.5.15] of S that $\widetilde{F'}^* F^* = Q\widetilde{F}' F Q$. The left-hand side of this equation has an analytic continuation into $\mathbb{C}^-$ and the right-hand side into $\mathbb{C}^+$. Thus the function $\widetilde{F}' F$ is an entire analytic function that approaches $\mathbb{1}$ at infinity. Hence by Liouville's theorem, $\widetilde{F}'(k)F(k) = \mathbb{1}$ for all $k \in \mathbb{C}^+$. In other words, F has a left inverse. This left inverse being a bounded operator, it follows that there can be no point $k_0 \in \mathbb{C}^+$ at which zero is in the continuous spectrum of $F(k_0)$.

The question now is: Is this left inverse also the right inverse? Suppose that for all k in some open set in $\mathbb{C}^+$, $F(k)$ has a right inverse, $F(k)F'_R(k) = \mathbb{1}$. Then in that set $\widetilde{F}'(k) = F'_R(k)$ and since $\widetilde{F}'$ is analytic, this equation has an analytic continuation into all of $\mathbb{C}^+$. Thus $\widetilde{F}'(k)$ is the *inverse* of $F(k)$ for all $k \in \mathbb{C}^+$.

Now $\|F(k) - \mathbb{1}\| \to 0$ as $|k| \to \infty$ in $\mathbb{C}^+$. It follows that $\|F^\dagger(k) - \mathbb{1}\| \to 0$ also. Therefore $\exists a \in \mathbb{R}_+$ such that for $|k| > a$ in $\mathbb{C}^+$, $F^\dagger(k)$ cannot have zero in its spectrum and hence it has a left inverse. Therefore $F(k)$ has a right inverse in an open set in $\mathbb{C}^+$. We thus have the following.

Lemma 2.6.2. Suppose that $W_0^1(S)$ and $W_0^1(S^*)$ both have solutions. Then for all $k \in \mathbb{C}^+$ every solution $F(k)$ of $W_0^1(S)$ has an inverse that is holomorphic in $\mathbb{C}^+$. This inverse is equal to $\widetilde{F}'(k)$, where F' solves $W_0^1(S^*)$.

We will also want to define a homogeneous version of W_0.
Problem $W_0^0(S)$: Let S be the operator defined as in Definition 1.5.15 in terms of a given operator $A \in \mathcal{A}$. Find a function F on $\mathbb{R}$ such that
  (i) $F \in \mathcal{N}^+$,
  (ii) on $\mathbb{R}$, F satisfies the equation
$$F^* = QS^* F Q.$$

In the case with bound states, on the other hand, we want the function F to have simple poles with residues that have given finite-dimensional ranges. (It is simpler to proceed in this way, and then to define the Jost function as the inverse of F.)
Problem $W_\sigma^1(S)$: Let S be the operator defined as in Definition 1.5.15 in terms of a given operator $A \in \mathcal{A}$. Let κ_m be a given' set of positive numbers and $\mathcal{H}_m$ a given set of finite-dimensional subspaces of $L^2(\mathbb{S}^2)$ in one-to-one correspondence with them, the sum of whose dimensions equals n_σ. We denote the set of positive numbers κ_m together with the corresponding spaces $\mathcal{H}_m$ by σ. Find a function F on $\mathbb{R}$ such that
  (i) $F - \mathbb{1} \in \mathcal{M}^+$, with simple poles at the points $i\kappa_m$ and residues there whose ranges equal $\mathcal{H}_m$;
  (ii) on $\mathbb{R}$, F satisfies the equation
$$F^* = QS^* F Q.$$

The homogeneous version of this problem, in which F is required to lie in $\mathcal{M}^+$, will be denoted by $W_\sigma^0(S)$.

Definition 2.6.3. For a given scattering amplitude $A \in \mathcal{A}_n$ the Jost function is defined as F^{-1} if that inverse exists everywhere in $\mathbb{C}^+$, where F solves the problem $W_\sigma^1(S)$ and σ is obtained from the scattering amplitude as explained in Section 1.5.

Thus the Jost function satisfies the equation

$$J^* = QJQS, \tag{2.77}$$

and its transpose solves $W_0^1(S^*)$. The definition of J and the manner in which $W_\sigma^1(S)$ is posed immediately imply the following.

Lemma 2.6.4. *If, for a given S matrix S, the Jost function J exists, then $\tilde{J} - 1 \in \mathcal{L}^2$ and $J - 1 \in \mathcal{N}^+$; it has zeros at the points $k = i\kappa_m$ such that nul $J(i\kappa_m) = \mathcal{H}_m$, where κ_m and $\mathcal{H}_m$ are the pairs defined in $W_\sigma^1(S)$. Its inverse solves $W_\sigma^1(S)$.*

The following results will be proved in Section 2.6.4.

Lemma 2.6.5. *(i) If W_σ^1 has a solution then this solution is unique if and only if W_σ^0 has only the trivial solution.*
(ii) If W_σ^1 has a unique solution then this solution is zero-free in $\mathbb{C}^+$. [By zero-free we mean that for no point k_0 is zero in the point spectrum of $F(k_0)$.]
(iii) If W_0^1 has a solution F that is zero-free in $\mathbb{C}^+$ and such that $F^{-1} - 1 \in \mathcal{M}^+$ then this solution is unique.
(iv) If $W_0^1(S)$ has a unique solution F and $F^{-1} - 1 \in \mathcal{M}^+$ then $W_0^1(S^*)$ has the unique solution $G = \tilde{F}^{-1}$ and F, G, $\tilde{F}$, and $\tilde{G}$ are all zero-free.
(v) If $W_\sigma^1(S)$ has a unique solution F and $F^{-1} - 1 \in \mathcal{M}^+$ then $F^{-1} - 1 \in \mathcal{N}^+$.

Corollary 2.6.6. *If the hypotheses of Lemma 2.6.2 are satisfied then the solutions F and F' of $W_0^1(S)$ and $W_0^1(S^*)$, respectively, are zero-free and unique.*

This follows from Lemma 2.6.5, (ii) and (iv).

The question arises if two different problems $W_\sigma^1(S)$ and $W_\mu^1(S)$, with $\sigma \neq \mu$ but the same S, can both have solutions. That that is possible if the two problems differ only in the position of their poles, but not in their number, is to be expected because that is the case for the "scalar" Riemann-Hilbert problem. We have the following two results, which will be proved in Section 2.6.4.

Lemma 2.6.7. *Let μ and σ be two sets as specified in the definition of problem $W_\sigma^1(S)$, such that $n_\mu = n_\sigma$. If $W_\sigma^1(S)$ has a unique solution F and $F^{-1} - 1 \in \mathcal{N}^+$, then $W_\mu^1(S)$ also has a unique solution with a holomorphic inverse. [See Notes]*

Theorem 2.6.8. (Index Theorem) *If $W_\sigma^1(S)$ and $W_\mu^1(S)$ have unique solutions and the inverses of these solutions are meromorphic in $\mathbb{C}^+$ then $n_\sigma = n_\mu$. This number will be called the* index *of S.*

This theorem shows that two different operator Riemann-Hilbert problems for the same symbol S cannot both have unique solutions with meromorphic inverses unless their indices are equal. Since the index is expected to be the sum of the dimensions of all the eigenspaces of the Schrödinger equation it is directly obtainable from S by means of the *Levinson Theorem*. This is the essential role of that theorem in the inverse problem.

2.6.1 The Reduction Method

There is a relatively simple method of reducing the solution of problem $W_\sigma^1(S)$ to another problem $W_0^1(S^{\text{red}})$ as follows.

Suppose first that σ contains only one pair; in other words, one is looking for a solution with one simple pole at $k = i\kappa$. One then constructs a real orthogonal projection $B = B^2 = \tilde{B} = \overline{B}$ onto the subspace $\mathscr{H}$, forms the function

$$\Pi_1(k) = \mathbb{1} - B + B\frac{k + i\kappa}{k - i\kappa}, \tag{2.78}$$

and defines a reduced S matrix,

$$S^{\text{red}} := Q\Pi_1^{-1}SQ\Pi_1^*. \tag{2.79}$$

The function S^{red} has all the essential properties needed in W_0^1. It satisfies the requirements of unitarity and reciprocity, i.e., (*ii*) and (*iv*) in the Definition 1.5.15 of $\mathscr{A}$; requirement (*iii*) is satisfied if B is chosen real, as it may because $\mathscr{H}$ is real; requirement (*vi*) will be ignored because we shall not use the reduction method for the actual solution of problem $W_\sigma^1(S)$ by means of a generalized Marchenko equation, but only to pose it schematically. A simple calculation shows that if S satisfies the Levinson theorem appropriate to one bound state eigenvalue of given degeneracy, then S^{red} satisfies the Levinson theorem appropriate to no bound states. Thus we proceed to solve $W_0^1(S^{\text{red}})$.

If there are more bound states then the above procedure has to be repeated for each of them. At each step, however, the space on which B projects has to be modified. Let Π_{m-1} be the operator constructed to remove the first $m-1$ poles. Then B_m is a projection onto $\Pi_{m-1}^{-1}(i\kappa_m)\mathscr{H}_m$ and

$$\Pi_m(k) := \Pi_{m-1}(k)\left[\mathbb{1} - B_m + B_m\frac{k + i\kappa_m}{k - i\kappa_m}\right]. \tag{2.80}$$

The reduced S matrix is then given by

$$S^{\text{red}} := Q\Pi_r^{-1}QS\Pi_r^*, \tag{2.81}$$

where r is the total number of bound states.

Suppose, then, that $W_0^1(S^{\text{red}})$ has a solution F^{red}; then the function $F := \Pi_r F^{\text{red}}$ solves the original problem $W_\sigma^1(S)$.

Theorem 2.6.9. $W_\sigma^1(S)$ *has a unique solution* F, *and this solution is such that* $F^{-1} - \mathbb{1} \in \mathscr{N}^+$, *if and only if* $W_0^1(S^{\text{red}})$ *has a unique solution with a meromorphic inverse. Here* S^{red} *is related to* S *by (2.81).*

Proof. Suppose that F' is the unique solution of $W_0^1(S^{\text{red}})$ and it has a mero-morphic inverse; by Lemma 2.6.5, item (iv), F^{-1} it is in fact holomorphic. Then $F := \Pi_r F'$ satisfies equation (2.76) and since F'^{-1} exists and is holomorphic in $\mathbb{C}^+$, F is meromorphic there and has the same poles as Π_r, with residues that have the same ranges. Furthermore, $F - \mathbb{1} \in \mathcal{M}^+$ and $F^{-1} - \mathbb{1} \in \mathcal{N}^+$, as the inverse of Π_r is holomorphic in $\mathbb{C}^+$. If there were other solutions of $W_\sigma^1(S)$ then by Lemma 2.6.5, item (i), $W_\sigma^0(S)$ would have a nontrivial solution and hence so would $W_0^0(S^{\text{red}})$. Thus F' would not be the unique solution of $W_0^1(S^{\text{red}})$, contrary to assumption.

Conversely, suppose that $W_\sigma^1(S)$ has a unique solution F and $F^{-1} - \mathbb{1} \in \mathcal{M}^+$. Then $F' := \Pi_r^{-1} F$ satisfies (2.76) with S^{red}, and Π_r being so designed that Π_r^{-1} annihilates the residues of F, F' has no poles and $F' - \mathbb{1} \in \mathcal{N}^+$. Thus it solves $W_0^1(S^{\text{red}})$. If there were other solutions, $W_0^0(S^{\text{red}})$ would have a nontrivial solution and hence so would $W_\sigma^0(S)$; thus, contrary to assumption, $W_\sigma^1(S)$ would not have a unique solution. Furthermore, $F'^{-1} = F^{-1}\Pi_r$ so that $F'^{-1} - \mathbb{1} \in \mathcal{M}^+$.

□

Although in principle the reduction method can be used to solve the problem $W_\sigma^1(S)$, it will turn out to be simpler to use another procedure given below.

2.6.2 Solution Procedure

Let us now turn to the actual solution of the operator Riemann-Hilbert problem. The method will be the same as that of Section 2.3 and we will make use of the machinery introduced there. We start with W_0^1.

Fourier transformation of (2.76) leads directly to the following two equations on $\mathbb{R}_+$:

$$\Gamma = GQ + \mathcal{G}\Gamma Q, \tag{2.82}$$
$$\Gamma = -QG^* - Q\mathcal{H}^*\Gamma, \tag{2.83}$$

where

$$\Gamma(\alpha) = \frac{1}{2\pi} \int_{-\infty}^{\infty} dk\, F(k) e^{-ik\alpha}, \tag{2.84}$$

and G, G^*, $\mathcal{G}$ and $\mathcal{G}^*$ are defined as before, except that now $x = 0$:

$$G(\alpha) = \frac{i}{(2\pi)^2} \int_{-\infty}^{\infty} dk\, kQA(k) e^{-ik\alpha},$$

$$\mathcal{G}(\alpha, \beta) := G(\alpha + \beta),$$
$$\mathcal{H}(\alpha, \beta) := G(\alpha - \beta),$$
$$\mathcal{G}^*(\alpha, \beta) := G(-\alpha - \beta),$$
$$\mathcal{H}^*(\alpha, \beta) := G(\beta - \alpha).$$

Define the following functions:

$$\Gamma_\pm := \tfrac{1}{2}\Gamma(\mathbb{1} \pm Q),$$
$$G_\pm := \tfrac{1}{2}G(\mathbb{1} \pm Q),$$

so that $\Gamma = \Gamma_+ + \Gamma_-$ and $G = G_+ + G_-$. Then (2.82) can be split into two equations:

$$(\mathbb{1} \mp \mathscr{G})\Gamma_\pm = \pm G_\pm. \tag{2.85}$$

The unitarity and reciprocity properties of the S matrix translate themselves into the following, which is proved by Fourier transformation just as was Lemma 2.3.2:

Lemma 2.6.10. *If for each $k \in \mathbb{R}$, $A(k)$ has properties (ii), (iii) and (iv) in Definition 1.5.15 then the following equations hold:*

$$(Q + \mathscr{H}^*)G_\pm = \mp(\mathbb{1} \pm \mathscr{G}^*)G_\pm^*. \tag{2.86}$$

As in the solution of the Riemann-Hilbert problem the aim is to solve W_0^1 by solving the generalized Marchenko equations (2.85) and avoid having to solve the Wiener-Hopf equation (2.83). Using Lemma 2.6.10 the proof of the following is quite analogous to that of Lemma 2.3.3 and will not be given.

Lemma 2.6.11. *If G and $\mathscr{G}$ are defined as above and $A \in \mathscr{A}$, then the equations (2.85), and hence (2.82), always have solutions in $\mathscr{L}^2$.*

Similarly, the following is proved as was Lemma 2.3.4:

Lemma 2.6.12. *If $\mathscr{G}^{*2}$ does not have the eigenvalue 1 then the right-handed Fourier transform*

$$F(k) = \mathbb{1} + \int_0^\infty d\alpha\, \Gamma(\alpha)e^{ik\alpha} \tag{2.87}$$

of any solution of the generalized Marchenko equation (2.82) solves the problem $W_0^1(S)$.

The fact that this lemma refers both to the eigenvalues $+1$ and -1 of the operator $\mathscr{G}^*$ originates, of course, from the presence of the two signs in (2.85).

It is clear that Lemma 2.3.5 holds directly for the homogeneous problem W_0^0 as well as to H_0^0. The result is the analogue of Theorem 2.3.6.

Theorem 2.6.13. *If $\mathscr{G}^{*2}$ does not have the eigenvalue 1 then the operator Riemann-Hilbert problem $W_0^1(S)$ has a solution and each such solution is obtained by right-handed Fourier transformation (2.87) from a solution of the generalized Marchenko equation (2.82). This solution is unique if and only if $\mathscr{G}^2$ does not have the eigenvalue 1.*

We now turn to the case with bound states. The Fourier transform Γ of F now no longer vanishes for negative values of its argument. Instead, since F has poles at the points $k = i\kappa_m \in \mathbb{C}^+$ with residues

$$\mathrm{Res}_m(\theta, \theta') = -i\sum_b Y_{\kappa_m}^b(\theta)\rho_m^b(\theta'),$$

the form of Γ for $\alpha < 0$ is given by

$$\Gamma(\alpha, \theta, \theta') = \sum_{m,b} y_m^b(-\alpha, \theta)\rho_m^b(\theta') := \mathscr{Y}(-\alpha, \theta, \theta'), \tag{2.88}$$

where the functions $Y^b_{\kappa_m}$ are the characters and the functions y^b_m are those defined in (2.35), except that now $x = 0$:

$$y^b_m(\alpha, \theta) = Y^b_{\kappa_m}(-\theta)e^{-\alpha\kappa_m};$$

the ρ^b_m are unknown coefficients, and b labels the degenerate bound states with the same eigenvalue $-\kappa^2_m$. Fourier transformation of (2.76) results in the following two sets of equations on $\mathbb{R}_+$:

$$(1 \mp \mathcal{G})\Gamma_\pm = \pm G_\pm \pm (Q + \mathcal{H})\mathcal{Y}_\pm, \tag{2.89}$$

$$(Q + \mathcal{H}^*)\Gamma_\pm = -G^*_\pm + (1 - \mathcal{G}^*)\mathcal{Y}_\pm, \tag{2.90}$$

where $\mathcal{Y}_\pm := 1/2\mathcal{Y}(1 \pm Q)$.

The functions ρ^b_m are obtained from (2.90) as was (2.42); for any $f \in L^2(\mathbb{S}^2)$

$$(\rho_\pm, f) = s^{-1}(z, G^*_\pm f), \tag{2.91}$$

where $(\cdot, \cdot)$ is the inner product on $L^2(\mathbb{S}^2)$, $\rho_\pm$ is the column matrix with entries $\rho^b_{m\pm} := \frac{1}{2}\rho^b_m(1 \pm Q)$, z is the column matrix with entries z_r (the functions that span the nullspace of $1 + \mathcal{G}^*$ as in Section 2.3), and s is the square matrix defined by (2.40). Insertion in (2.89) and (2.90) leads to the equations

$$\Gamma_\pm = \pm G_\pm \pm (Q + \mathcal{H})\Lambda G^*_\pm \pm \mathcal{G}\Gamma_\pm, \tag{2.92}$$

$$(Q + \mathcal{H}^*)\Gamma_\pm = -PG^*_\pm, \tag{2.93}$$

where Λ is defined in (2.43) and P is the projection (2.44).

These equations have the same structure as equations (2.45) and (2.46). The proof of the following is entirely parallel to that of Lemma 2.3.10.

Lemma 2.6.14. *If the functions in (2.92) are as previously defined, that equation always has a solution in $\mathcal{L}^2$, even if its homogeneous version has a nontrivial solution.*

In order to solve the problem $W^1_\sigma(S)$ by means of solving (2.92), the Fourier transform (2.84) has to be inverted, using (2.88):

$$F(k, \theta, \theta') = \int_0^\infty d\alpha\, e^{ik\alpha}\Gamma(\alpha, \theta, \theta') + \frac{1}{i(k - i\kappa_m)}\sum_{m,b} Y^b_{\kappa_m}(\theta)\rho^b_m(\theta'). \tag{2.94}$$

The following is the analogue of Theorem 2.3.11 and the extension of Theorem 2.6.13. [The argument used in the proof of Theorem 2.3.11 is modified by employing (2.86) in place of (2.30).]

Theorem 2.6.15. *The problem $W^1_\sigma(S)$ with the index n_σ has a unique solution if and only if the operator $\mathcal{G}^2$ does not have the eigenvalue 1, $\dim \mathrm{nul}(1 + \mathcal{G}^*) = n_\sigma$, and the matrix s of (2.40) is invertible. The solution is obtained by (2.94) from the solution of the generalized Marchenko equation (2.92).*

Remark. Forward analyticity of the scattering amplitude, and the resulting connection between the scattering amplitude and the bound states, has not been taken into account in the above statements. This additional constraint arises from the existence of an underlying potential, which has not been an issue in the entire discussion of this section.

Let us now look at the Jost function of a shifted potential. If the operator Riemann-Hilbert problems are set up for S_x rather than S then, of course, the solution will depend on x. Comparison of Problem $W_\sigma^1(S_x)$ with Problem $H_\sigma^1(S_x)$ shows that if F_x solves $W_\sigma^1(S_x)$ then $F_x\hat{1}$ solves $H_\sigma^1(S_x)$, as $Q\hat{1} = \hat{1}$. Therefore the solution of the Riemann-Hilbert problem H_σ^1 may be subsumed under that of W_σ^1. Specifically, $\mathscr{H}_m$ is the character space of the eigenvalue $-\kappa_m^2$ and n_σ is the sum of the dimensions of the eigenspaces.

Since the function $F_x\hat{1}$ satisfies the modified Schrödinger equation (1.53) one is led to inquire if F_x itself satisfies any kind of differential equation with respect to x. This question can be answered as in Section 2.4.

Apply the differential operator $(\Delta - 2\frac{\partial}{\partial\alpha}\theta \cdot \nabla)$ to equations (2.92) and (2.93), regarding θ as a multiplicative operator, and assume that the operator defined by

$$V := 2\theta \cdot \nabla[\Gamma(\alpha = 0-) - \Gamma(\alpha = 0+)]$$

is such that $QVQ = V$. (This is guaranteed to be the case if $\theta \cdot \nabla G$ is continuous at $\alpha = 0$, which, however, is not assured.) One then finds that $(\Delta - 2\frac{\partial}{\partial\alpha}\theta \cdot \nabla)\Gamma - \Gamma V$ satisfies the homogeneous version of (2.92) and (2.93). Therefore, if these homogeneous equations have only the trivial solution, i.e., if $\mathscr{G}^2$ does not have the eigenvalue 1, then the function $\Gamma(\alpha, \theta, \theta', x)$ satisfies the equation

$$\left(\Delta - 2\frac{\partial}{\partial\alpha}\theta \cdot \nabla\right)\Gamma(\alpha, \theta, \theta', x) = \int_{S^2} d\theta'' \, \Gamma(\alpha, \theta, \theta'', x)V(\theta'', \theta', x),$$

and thus its Fourier transform satisfies a Schrödinger equation. (Here $V(\theta'', \theta', x)$ is the integral kernel of the operator V.) However, the meaning or implication of this equation with the operator potential V in it, is unclear. The miracle, from this point of view, is that V, for all x, has $\hat{1}$ as an eigenfunction; the corresponding eigenvalue is $V(x)$.

2.6.3 Angular Momentum Projections

For later purposes we will need the angular-momentum projections of the transpose of the Jost function, which are defined like those of the scattering amplitude in Section 1.6. We define

$$\mathscr{L}(k) := \tilde{J}(k) - \mathbb{1}$$

and its angular-momentum projections

$$\mathscr{L}_{LL'}(k) := \int_{S^2 \times S^2} d\theta d\theta' \, Y_L(\theta)\overline{Y_{L'}(\theta')}\mathscr{L}(k, \theta, \theta').$$

We then have the following result. [The functions Y_L here are defined as in Section 1.6.]

Lemma 2.6.16. *If the scattering amplitude $A(k)$ is such that the conclusion of Lemma 1.6.1 holds then, generically, the Jost function is such that near $k = 0$*

$$\mathscr{L}_{L'L}(k) = O(k^n),$$

where $n = \max(0, l - l')$. Exceptions to this behavior may be possible.

Proof. We define the Fourier transforms

$$\Lambda(\alpha) := \frac{1}{2\pi} \int_{-\infty}^{\infty} dk \, e^{-ik\alpha} \mathscr{L}(k),$$

$$\Lambda_{LL'}(\alpha) := \frac{1}{2\pi} \int_{-\infty}^{\infty} dk \, e^{-ik\alpha} \mathscr{L}_{LL'}(k).$$

The Fourier transform of the transpose of (2.77) then becomes

$$\mathscr{L}(\alpha)Q - Q\mathscr{L}(-\alpha) = \mathsf{G}(-\alpha) + \int_{-\infty}^{\infty} d\beta \, \mathsf{G}(-\alpha - \beta)\mathscr{L}(\beta),$$

expansion of which on the basis of the spherical harmonics reads

$$(-1)^l \Lambda_{L'L}(\alpha) - (-1)^{l'} \Lambda_{L'L}(-\alpha) = G_{L'L}(-\alpha)$$

$$+ \sum_{L''} \int_{-\infty}^{\infty} d\beta \, G_{L'L''}(-\alpha - \beta)\Lambda_{L''L'}(\beta), \qquad (2.95)$$

where $G_{LL'}$ is defined in (1.95). We multiply this equation by α^n, integrate with a shift of the variable of integration, and use the binomial theorem, defining

$$\Gamma^n_{L'L} := \frac{1}{n!} \int_{-\infty}^{\infty} d\alpha \, \alpha^n \Lambda_{L'L}(\alpha).$$

The result is the equation

$$[(-1)^{l+n} - (-1)^{l'}]\Gamma^n_{L'L} = \rho^n_{L'L} + \sum_{l''=0}^{n-l'-1} \sum_{m''=-l''}^{l''} \sum_{s=0}^{n-l'-l''-1} \rho^{n-s}_{L'L''}\Gamma^s_{L''L}. \qquad (2.96)$$

The upper limits on the sums originate from Corollary 1.6.3. For each fixed n and L, (2.96) constitues a finite set of equations for $\Gamma^s_{L'L}$ with $s \le n$ and $l' \le n$. Consider the set of all $\Gamma^n_{L'L}$ with $l' + n < l$. This is a finite set, and all its members satisfy the homogeneous form of (2.96). While there may exist exceptional sets of $\rho^n_{LL'}$ for which these equations do not imply that the solutions vanish, in the generic case they lead to $\Gamma^n_{L'L} = 0$ for all $n < l - l'$. The lemma now follows by inverting the Fourier transform. $\qquad \square$

Remark. With $t = n - l'$ equation (2.96) becomes for $t \ge l$, $t - l$ even,

$$\sum_{l''=0}^{t-1} \sum_{m''=-l''}^{l''} \sum_{s=0}^{t-l''-1} \rho^{t+l'-s}_{L'L''} \Gamma^s_{L''L} = -\rho^{t+l'}_{L'L}.$$

For fixed t and l this is an infinite set of inhomogeneous equations for the finite set of $\Gamma^s_{L''L}$. Thus, unless the $\rho^n_{L'L}$ are, again, exceptional there will be no solution. It follows that in general, even if the scattering amplitude is analytic at $k = 0$ and satisfies the conclusion of Lemma 1.6.1, the Jost function will not be analytic at $k = 0$.

2.6.4 Proofs

Proof of Lemma 2.6.5. The proof of (i) is obvious.

(ii) Suppose that f solves W_σ^1 and has a zero at $k = \kappa$, $\kappa \in \mathbb{C}^+$; let P be a projection on the nullspace nul $f(\kappa)$. Then the function defined by $f' := f[P1/(k-\kappa)-QPQ1/(k+\kappa)]$ solves W_σ^0, and hence by Lemma 2.6.5 the solution of W_0^1 is not unique.

(iii) Suppose that f is zero-free in $\mathbb{C}^+$ and that $f - 1 \in \mathcal{M}^+$. Then, in fact, $f - 1 \in \mathcal{N}^+$. Let g be another solution of W_0^1; then $f^{*-1}g^* - 1 = f^{-1}g - 1$, so that both sides of this equation are entire analytic functions that vanish at infinity. Therefore by Liouville's theorem they vanish identically, $f^{-1}g = 1$. Now, if f^{-1} had a zero in $\mathbb{C}^+$ then $f - 1$ could not be in $\mathcal{N}^+$. Therefore f^{-1} is zero-free and $f = g$.

(iv) By (ii) F is zero-free and since by assumption $F^{-1} - 1 \in \mathcal{M}^+$, in fact $F^{-1} - 1 \in \mathcal{N}^+$ and hence $\widetilde{F}$ is also zero-free. It follows from (2.76) that on $\mathbb{R}$ $\widetilde{F}^{-1*} = QS\widetilde{F}^{-1}Q$. Thus $G = \widetilde{F}^{-1}$ solves $W_0^1(S^*)$. Since F and $\widetilde{F}$ are holomorphic, G and $\widetilde{G}$ are zero-free, and by (iii) G is the unique solution of $W_0^1(S^*)$.

(v) By Theorem 2.6.9 $W_\sigma^1(S)$ has a unique solution F whose inverse is meromorphic in $\mathbb{C}^+$ if and only if $W_0^1(S^{\text{red}})$ has a unique solution F' whose inverse is meromorphic. But by Lemma 2.6.5, item (iv), the latter is then, in fact, holomorphic and since Π_r^{-1} has no poles in $\mathbb{C}^+$, $F^{-1} = F'^{-1}\Pi_r^{-1}$ is also holomorphic in $\mathbb{C}^+$. $\qquad\square$

Proof of Lemma 2.6.7. Assume thet F solves $W_\sigma^1(S)$, in which one of the poles is stipulated to be at $k = i\kappa$, and let the residue of F there be R. Define $F_\pm := 1/2F(1 \pm Q)$, so that $F_\pm^* = \pm QS^*F_\pm$, and $F = F_+ + F_-$; define also $F_\pm' := F_\pm \Pi_\pm$,

$$\Pi_\pm^{-1} := 1 + C_\pm \frac{v^2 - \kappa^2}{k^2 + \kappa^2}.$$

Let $R_\pm$ be the residue of $F_\pm$ at $k = i\kappa$ and take $C_\pm$ so that $R_\pm(1-C_\pm) = 0$. Then $F_\pm'$ are holomorphic at $k = i\kappa$. Since $R_\pm = R(1\pm Q)/2$, we define $C_\pm$ so that the range of $1-C_\pm$ equals the nullspace of $R_\pm$, and $C_\pm = (1\pm Q)C_\pm/2$; we also choose $C_\pm$ to be projections, $C_\pm^2 = C_\pm$, and so that $C_\pm = C_\pm(1\pm Q)/2$, which is always possible. This implies that $C_+C_- = C_-C_+ = 0$ and $C := C_+ + C_-$ is also a projection. We then define $F' := F_+' + F_-'$ and find $F'^* = F_+'^* + F_-'^* = F_+^*\Pi_+ + F_-^*\Pi_- = QS^*(F_+\Pi_+Q - F_-\Pi_-Q)Q$, and one easily sees that $F_+\Pi_+Q - F_-\Pi_-Q = F'$. Therefore F' satisfies (2.76).

The functions $F_\pm'$ have poles at $k = iv$ and their residues there are $R_\pm' = F_\pm(iv)C_\pm$. Thus F' has a pole there with residue $R' = R_+' + R_-' = F_+(iv)C_+ + F_-(iv)C_- = F(iv)[(1+Q)C_+/2 + (1-Q)C_-/2] = F(iv)C$. Thus if $F(k)$ has no zero at $k = iv$, the dimensionality of the range of R' equals that of the range of C.

Since nul $R = $ ran $(1-C) = $ nul C and C is a projection, nul $R \cap$ ran $C = \{0\}$. It follows that $R = RC$ implies that the range of C has the same dimensionality as the range of R. Therefore dim ran $R = $ dim ran $C = $ dim ran R'. Thus we conclude that F' is a solution of $W_\mu^1(S)$ in which one of the poles has been shifted to a new position, but $n_\mu = n_\sigma$. The range of the residue at the new pole can be arbitrarily assigned by proper choice of $C_\pm$ if $F(iv)$ has a left inverse, which by Lemma 2.6.2 it does. This can be repeated for the other poles.

If we choose $\operatorname{nul} C = \operatorname{ran}(1 - C) \supset \operatorname{nul} R$ with $m := \dim \operatorname{ran} C < \dim \operatorname{ran} R$, then F still has a pole at $k = i\kappa$ with a residue $R(1 - C)$ such that $\dim \operatorname{ran} R(1 - C) = \dim \operatorname{ran} R - m$ and it also has a simple pole at $k = iv$ with residue $F(iv)C$ such that $\dim \operatorname{ran} F(iv)C = m$. Had we chosen $\dim \operatorname{ran} C > \dim \operatorname{ran} R$ then F^{-1} would have a pole at $k = i\kappa$, which we do not want.

In a similar manner one can combine two poles into one with a residue the dimension of whose range is the sum of the old, and the range of a residue may be changed without a change in its dimension or in the pole position by performing two shifts. [See Notes] □

Proof of Theorem 2.6.8. By item (v) of Lemma 2.6.5 the unique solutions F_1 and F_2, of W_σ^1 and W_μ^1, respectively, have inverses that are holomorphic in $\mathbb{C}^+$. Define $\chi := F_1^{-1} F_2$. Then from (2.76) $\chi^* = Q\chi Q$ and thus χ is meromorphic in $\mathbb{C}$, and so is its inverse. The poles of χ in $\mathbb{C}^+$ are those of F_2, in $\mathbb{C}^-$ those of F_2^*, and the poles of χ^{-1} are those of F_1 and F_1^*, respectively. If any of the zeros of F_1^{-1} (or of F_1^{*-1}) coincide with poles of F_2 (or of F_2^*), we shift them by the method used in the proof of Lemma 2.6.7.

Assuming, then, that no pole positions of χ coincide with zeros of F_1^{-1} or F_1^{*-1}, the dimensions of the ranges of the residues of the poles of χ and of χ^{-1} equal those of the corresponding ones of F_2, F_2^*, F_1, and F_1^*, respectively. We now remove the poles of χ one by one. Without loss of generality we may assume that the sum of the dimensions of the nullspaces at the zeros of χ are not smaller than the sum of the dimensions of the ranges of the residues at is poles. Let λ be a zero of χ and κ a simple pole with residue R. Let $B^2 = B$ be a projection on a one-dimensional subspace (spanned by η) of $\operatorname{ran} R$ such that $B\chi(\lambda) = 0$. Such a projection exists, provided that $\operatorname{ran} R \not\subset \operatorname{ran} \chi(\lambda)$. If necessary we use the method given in the proof of Lemma 2.6.7 to change $\operatorname{ran} R$. Define

$$\Pi := 1 + B\left(\frac{\lambda - \kappa}{k - \lambda}\right), \qquad \chi' := \Pi \chi.$$

Since $B\chi(\lambda) = 0$, χ' is analytic at λ. At $k = \kappa$, χ' has a simple pole with residue $R' = (1 - B)R$. Therefore $\operatorname{ran} R' \subset \operatorname{ran} R$. Furthermore, if there exists an a such that $\eta = R'a$ then $\eta = R'a = (1 - B)Ra$, and hence $B\eta = 0$; and if $\eta = Ra$ then $R'a = (1 - B)Ra = (1 - B)\eta = 0$. Thus η is not in $\operatorname{ran} R'$, and $\dim \operatorname{ran} R' = \dim \operatorname{ran} R - 1$. Since $\chi'(\lambda) = \chi(\lambda) + (\lambda - \kappa)B\hat\chi(\lambda)$, where $\hat\chi = d\chi/dk$, and $B\chi(\lambda) = 0$, $\chi'(\lambda)a = 0$ implies $\chi(\lambda)a = 0$ and $B\chi'(\lambda) = 0$. Thus $\operatorname{nul} \chi' \subset \operatorname{nul} \chi$ and $\operatorname{ran} \chi' = \operatorname{ran} \chi \cup \{\eta\}$, from which it follows that $\dim \operatorname{nul} \chi'(\lambda) = \dim \operatorname{nul} \chi(\lambda) - 1$. This procedure is repeated until all the poles of χ have been removed. By Liouville's theorem the remaining factor must then be equal to 1. In each step the dimensions of the ranges of the residues of χ and of its nullspaces have been lowered by one. Therefore the sums of these dimensions are equal. □

2.7 Perturbations and Stability

In this section we shall modify the generalized Marchenko equation so as to solve problems in which the starting point is not a "free" Schrödinger equation with plane-wave solutions but instead, a Schrödinger equation with a known potential in it. It will be assumed that for this "unperturbed" potential V_0 everything is known, including the Jost function J_{0x} for a shifted potential and its inverse $F_{0x} = J_{0x}^{-1}$. Specifically this includes the assumption that this Jost function and its inverse exist, which was seen in the last section not to be completely assured. It will also be assumed for simplicity that neither the unperturbed potential V_0 nor the perturbed potential V cause bound states. In order not to clutter up the notation we shall not explicitly show the x-dependence of the functions involved until that becomes necessary. It will be understood that everything refers to shifted potentials.

The function F_0 satisfies the equation

$$F_0^* = QS_0^* F_0 Q,$$

while $\zeta = F\hat{1}$ satisfies

$$\zeta^* = QS^*\zeta.$$

As a result one obtains

$$
\begin{aligned}
J_0^*\zeta^* - \hat{1} &= J_0^* Q(S^* - S_0^*)F_0(J_0\zeta - \hat{1}) \\
&\quad + Q(J_0\zeta - \hat{1}) + J_0^* Q(S^* - S_0^*)F_0\hat{1}.
\end{aligned}
$$

If we define

$$\mathscr{F} := J_0\zeta - \hat{1}, \qquad \Omega := QJ_0Q(S - S_0)F_0^*,$$

then this equation reads

$$\mathscr{F}^* = Q\mathscr{F} + Q\Omega^*\hat{1} + Q\Omega^*\mathscr{F}.$$

We take its Fourier transform after defining

$$\eta_1(\alpha) := \frac{1}{2\pi} \int_{-\infty}^{\infty} dk\, \mathscr{F}(k)e^{-ik\alpha},$$

$$G_1(\alpha) := \frac{1}{2\pi} \int_{-\infty}^{\infty} dk\, Q\Omega(k)e^{-ik\alpha}, \qquad g_1(\alpha) := G_1(\alpha)\hat{1}.$$

The result is

$$\eta_1(\alpha) = Q\eta_1(-\alpha) + g_1(\alpha) + \int_0^{\infty} d\beta\, G_1(\alpha + \beta)\eta_1(\beta),$$

because the analyticity of $\mathscr{F}$ in $\mathbb{C}^+$ implies that $\eta_1(\alpha) = 0$ for $\alpha < 0$. Thus we obtain the two equations for $\alpha > 0$:

$$\eta_1 = g_1 + \mathscr{G}_1\eta_1, \tag{2.97}$$

$$\eta_1 = -Qg_1^* - Q\mathscr{H}_1^*\eta_1, \tag{2.98}$$

where $\mathscr{G}_1$ and $\mathscr{H}_1^*$ are the operators defined by

$$(\mathscr{G}_1 f)(\alpha) = \int_0^\infty d\beta \, G_1(\alpha + \beta) f(\beta),$$

$$(\mathscr{H}_1^* f)(\alpha) = \int_0^\infty d\beta \, G_1(\beta - \alpha) f(\beta)$$

for $\alpha > 0$.

The connection with the potential is made as follows. If the functions η and η_0 are defined as before, namely, the Fourier transforms of the functions $\zeta - \hat{1} = F\hat{1} - \hat{1}$ and $\zeta_0 - \hat{1} = F_0\hat{1} - \hat{1}$ then $\zeta - \zeta_0 = F_0\mathscr{F}\hat{1}$ and we have by Fourier transformation

$$\eta_1(\alpha) = \eta(\alpha) - \eta_0(\alpha) - \int_0^\alpha d\beta \, \Gamma_0(\alpha - \beta) \eta_1(\beta)$$

for $\alpha > 0$, where

$$\Gamma_0(\alpha) := \frac{1}{2\pi} \int_{-\infty}^\infty dk \, [F_0(k) - \mathbb{1}] e^{-ik\alpha}.$$

Therefore,

$$\begin{aligned}
\eta_1(0+, \theta, x) &= \eta(0+, \theta, x) - \eta_0(0+, \theta, x) \\
&= -\frac{1}{2} \int_0^\infty dr \, [V(x - \theta r) - V_0(x - \theta r)]
\end{aligned} \tag{2.99}$$

if S is admissible. It follows that if S is admissible then the solution of (2.97) must be such that $\theta \cdot \nabla \eta_1(\alpha = 0+, \theta, x)$ is independent of θ; that is to say, η_1 must be miraculous. Conversely, if η_1 is miraculous, then so is η and the potential exists.

In order to test the stability of the solution of the generalized Marchenko equation we investigate what happens if S is close to S_0.

Lemma 2.7.1.

$$\|\mathscr{G}_1\| \leq \frac{1}{2\pi} \sup_{-\infty < k < \infty} \|k J_0(k) Q[A(k) - A_0(k)] F_0(-k)\|_{S^2}. \tag{2.100}$$

The norm on the left is the operator norm on $L^2(\mathbb{R}_+ \times S^2)$ and that on the right is the operator norm on $L^2(S^2)$.

Proof. Define, for $\alpha > 0$,

$$\eta(\alpha) := \int_0^\infty d\beta \, G_1(\alpha + \beta) \xi(\beta),$$

where $\xi(\beta) = 0$ for $\beta < 0$ and $\xi \in L^2(\mathbb{R}_+ \times S^2)$,

$$\xi(\beta) := \frac{1}{2\pi} \int_{-\infty}^\infty dk \, f(k) e^{-ik\beta}.$$

Then

$$\eta(\alpha) = \frac{1}{2\pi} \int_{-\infty}^{\infty} dk\, Q\Omega(k) f(-k) e^{-ik\alpha}$$

$$= \frac{1}{2\pi} \int_{-\infty}^{\infty} dk\, g(k) e^{-ik\alpha},$$

$$g(k) = Q\Omega(k) f(-k).$$

Hence,

$$\|\eta\|_+^2 := \int_0^{\infty} d\alpha\, \|\eta(\alpha)\|_{\mathfrak{S}^2}^2 \leq \int_{-\infty}^{\infty} d\alpha\, \|\eta(\alpha)\|_{\mathfrak{S}^2}^2 = \frac{1}{2\pi} \int_{-\infty}^{\infty} dk\, \|g(k)\|_{\mathfrak{S}^2}^2,$$

which implies that

$$\|\mathscr{G}_1 \xi\|_+^2 \leq \frac{1}{2\pi} \int_{-\infty}^{\infty} dk\, \|\Omega(k) f(-k)\|_{\mathfrak{S}^2}^2$$

$$= \frac{1}{(2\pi)^2} \int_{-\infty}^{\infty} dk\, k^2 \|J_0(k) Q[A(k) - A_0(k)] F_0(-k) f(-k)\|_{\mathfrak{S}^2}^2$$

$$\leq \sup_{-\infty < k < \infty} k^2 \|J_0(k) Q[A(k) - A_0(k)] F_0(-k)\|_{\mathfrak{S}^2}^2 \frac{1}{2\pi} \|\xi\|^2. \quad \square$$

Corollary 2.7.2. $\|\mathscr{G}_1\|$ *can be made arbitrarily small by choosing A and A_0 sufficiently close to one another.*

Now it follows from (2.97) that

$$\eta_1 = (\mathbb{1} - \mathscr{G}_1)^{-1} g_1 = g_1 + (\mathbb{1} - \mathscr{G}_1)^{-1} \mathscr{G}_1 g_1,$$

and therefore, if A and A_0 are close enough in the sense of Lemma 2.7.1, then $\eta_1 \simeq g_1$. As a consequence we have, from the above results, the following.

Theorem 2.7.3. *If A and A_0 are sufficiently close to one another in the sense of Lemma 2.7.1, then*

$$\int_0^{\infty} dr [V(x - \theta r) - V_0(x - \theta r)]$$

$$\simeq \frac{1}{2i\pi^2} \int_{-\infty}^{\infty} dk\, k J_{0x}(k) Q[A_x(k) - A_{0x}(k)] \zeta_{0x}(-k). \tag{2.101}$$

(For the sake of clarity we have explicitly indicated the x-dependence.) This result may also be expressed in the form of a Frechet derivative:

$$\frac{\delta \int_0^{\infty} dr\, V(x - \theta r)}{\delta A(k, \theta_1, \theta_2)} = \frac{k}{2i\pi^2} J_x(k, \theta, -\theta_1) \psi^+(-k, \theta_2, x) e^{ik\theta_1 \cdot x}. \tag{2.102}$$

Note that the error-term neglected in (2.101) by virtue of (2.100) is small in an "average" sense and not pointwise in θ. Note also that there is no guarantee that $J_{0x}(k)$ and $F_{0x}(k)$ exist for each k, even if they exist as operator-valued L^2-functions. It is therefore inadvisable to take these factors out of the operator norm on the right-hand side of (2.100).

The results (2.101) and (2.102) provide the mapping $A \mapsto V$ with a certain measure of smoothness. The potential itself is obtained by differentiating with respect to x:

$$V(x) - V_0(x) \simeq \frac{1}{2i\pi^2} \theta \cdot \nabla \int_{-\infty}^{\infty} dk \, k J_{0x}(k) Q[A_x(k) - A_{0x}(k)] \zeta_{0x}^{\bullet}, \tag{2.103}$$

and similarly

$$\frac{\delta V(x)}{\delta A(k, \theta_1, \theta_2)} = \frac{k}{2i\pi^2} \theta \cdot \nabla [J_x(k, \theta, -\theta_1) \psi^+(-k, \theta_2, x) e^{ik\theta_1 \cdot x}]. \tag{2.104}$$

These equations make sense only if their right-hand sides are miraculous, that is, if they are independent of θ. They do not tell us how to distort an admissible scattering amplitude so as to obtain, again, an admissible scattering amplitude.

On the other hand, suppose that A_0 is admissible but A is not. Then the function V in (2.103), or that calculated from A by means of the generalized Marchenko equation, is not independent of θ and there is no potential that corresponds to A. Nevertheless, (2.103) and (2.104) hold and ensure that if the inadmissible A differs from an admissible A_0 by a small amount, in the sense of (2.101) and (2.103), then the "pseudo-potential" computed from A by means of the generalized Marchenko equation differs from the potential V_0 that corresponds to A_0 also by a small amount. This means that the inversion procedure has the minimal kind of stability needed for application to experimental data. Since such data inevitably include errors, no experimentally given function A can be expected to be admissible. However, if there is an admissible scattering amplitude A_0 nearby, then the "pseudo-potential" computed from A is close to the potential V_0 that corresponds to A_0.

One may worry that some admissible scattering amplitudes are isolated, i.e., that for some admissible A_0 there is no nearby A for which the right-hand side of (2.103) is independent of θ. This possibility is ruled out by the results of perturbation theory in the direct scattering problem. It is well known that the scattering amplitude is a continuous function of the potential. In fact, [Ne82c, p. 239]

$$\delta A(k, \theta, \theta') = -\frac{1}{4\pi} \int_{\mathbb{R}^3} dx \, \delta V(x) \psi^+(k, -\theta, x) \psi^+(k, \theta', x).$$

This shows that in any neighborhood of an admissible scattering amplitude there is another admissible scattering amplitude. Thus the admissible set cannot contain isolated points.

The arguments of this section break down only if the Jost function or its inverse fails to exist. In that case Lemma 2.7.1 and Theorem 2.7.3 fail, and the mapping $A \mapsto V$ may not be continuous.

2.8 A Representation of the Potential

The results of the generalized Marchenko method can be used to establish a representation of the potential that may have a variety of uses. Let us employ the notation introduced in equations (1.50) and write equation (2.1) in the form

$$\zeta(-k, \theta, x) = \zeta(k, \theta, x) + \frac{k}{2\pi i} \int_{S^2} d\theta' \, A(-k, \theta, \theta') v(k, \theta, \theta', x),$$

the Fourier transform of which reads

$$\eta(\alpha, -\theta, x) - \eta(-\alpha, \theta, x)$$
$$= \frac{-i}{(2\pi)^2} \int_{-\infty}^{\infty} dk\, k \int_{S^2} d\theta'\, e^{ik\alpha} A(-k, \theta, \theta') v(k, \theta, \theta', x).$$

The use of equations (2.55) and (2.36) then leads to the following representation for the potential that underlies the given scattering amplitude

$$V(x) = \theta \cdot \nabla \left\{ \frac{-i}{2\pi^2} \int_{-\infty}^{\infty} dk\, k \int_{S^2} d\theta'\, A(-k, \theta, \theta') e^{-ik\theta \cdot x} \psi^+(k, \theta', x) \right.$$
$$\left. + \sum_{n,b} \kappa_n^{-1} u_n^b(x) [Y_n^b(-\theta) e^{\kappa_n \theta \cdot x} - Y_n^b(\theta) e^{-\kappa_n \theta \cdot x}] \right\}, \tag{2.105}$$

where $u_n^b(x)$ are orthonormal eigenfunctions of the corresponding Schrödinger equation with the eigenvalues $-\kappa_n^2$, and $\psi^+(k, \theta, x)$ are the scattering solutions. This equation may be regarded as the inversion of the representation (1.62) of the scattering amplitude in terms of the potential. In fact, (2.105) can alternatively be obtained directly from (1.62) by means of the formula (1.51). Unfortunately, since the functions v' defined in (1.50) have not been proved to be linearly indendent, it is not known whether the function $A(k, \theta, \theta')$ that appears in the right-hand side of (2.105) is necessarily identical with the scattering amplitude that corresponds to the potential on the left. In other words, although the potential has the representation (2.105) in terms of its scattering amplitude and bound states, it has not been proved that it may not have the same representation in terms of another function.

The representation (2.105) may also be cast into another form. The version of the Schrödinger equation for the function v used in (2.105) leads to the equation

$$(\Delta - V)\Gamma = 0, \tag{2.106}$$

where

$$\Gamma(x) : = 1 + \frac{1}{(2\pi)^2} \int_{-\infty}^{\infty} dk \int_{S^2} d\theta'\, A(-k, \theta, \theta') v(k, \theta, \theta', x)$$
$$+ \sum_{n,b} \frac{1}{2\kappa_n^2} u_n^b(x) \left[Y_n^b(\theta) e^{-\kappa_n \theta \cdot x} + Y_n^b(-\theta) e^{\kappa_n \theta \cdot x} \right]. \tag{2.107}$$

As $|x| \to \infty$, $\Gamma(x) \to 1$; therefore Γ is the scattering solution for $k \to 0$, $\Gamma(x) = \psi^+(0, \theta, x)$, which is independent of θ. Defining $\Lambda(x) := 1/\Gamma(x)$ and $\varphi(k, \theta, x) := \psi^+(k, \theta, x)\Lambda(x)$, $v_n^b(x) := u_n^b(x)\Lambda(x)$, one obtains

$$\Lambda(x) = 1 - \frac{1}{(2\pi)^2} \int_{-\infty}^{\infty} dk \int_{S^2} d\theta'\, A(-k, \theta, \theta') e^{-ik\theta \cdot x} \varphi(k, \theta', x)$$
$$- \sum_{n,b} \frac{1}{2\kappa_n^2} v_n^b(x) \left[Y_n^b(\theta) e^{-\kappa_n \theta \cdot x} + Y_n^b(-\theta) e^{\kappa_n \theta \cdot x} \right], \tag{2.108}$$

and the Lippmann-Schwinger equation for φ reads

$$\varphi(k, \theta, x) = e^{ik\theta \cdot x} - \frac{1}{2\pi} \int_{\mathbb{R}^3} dy\, \frac{e^{ik|x-y|}}{|x - y|} [\nabla \log \Lambda(y)] \cdot \nabla \varphi(k, \theta, y). \tag{2.109}$$

The nonlinear system (2.108) and (2.109) may be regarded as an alternative basis for solving the inverse scattering problem. Note that the data needed here consist only of the scattering amplitude with either the incoming or the outgoing direction fixed. Since that is a function of three variables, this method would not be intrinsically redundant. Nothing is known about the solvability of this system, the existence or uniqueness of solutions to it, or whether the potential obtained from it necessarily has the input function $A(k, \theta, \theta')$ as its scattering amplitude.

There is a simple consequence of the fact that the function $\Gamma(x)$ is equal to the scattering solution of the Schrödinger equation for $k = 0$, and therefore $\int_{\mathbb{R}^3} dx\, V(x)\Gamma(x) = -4\pi A(0)$. The result of multiplying (2.107) by $V(x)$ and integrating is, by equation (1.29),

$$\int_0^\infty dk\, \sigma(k, \theta) = \frac{\pi}{2} \int_{\mathbb{R}^3} dx\, V(x) + 2\pi^2 A(0) + \frac{\pi}{2} \sum_{n,b} \frac{1}{\kappa_n^2} Y_n^b(\theta) Y_n^b(-\theta),$$

in which $A(0)$ is the scattering amplitude at $k = 0$, which is independent of θ and θ', and $\sigma(k, \theta) = \int_{S^2} d\theta' \, |A(k, \theta', \theta)|^2$ is the total cross section for scattering from the incident direction θ. This equation shows that its left-hand side must be independent of θ if there are no bound states. On the other hand, if V produces bound states then the sum on the right-hand side will generally depend on θ and hence so will the left-hand side. For central potentials the character functions are simply spherical harmonics and the sum on the right does not depend on θ, and neither does the left-hand side.

2.9 Miscellaneous Methods

There are two other inversion methods that will not be treated in detail, but which deserve mention.

The Moses-Prosser System. If equation (1.30) is inserted in formula (1.62) for the scattering amplitude, the result is

$$-4\pi A(|k|, \hat{k}', \hat{k}) = \widehat{V}(k - k')$$
$$+ \frac{1}{(2\pi)^3} \int_{\mathbb{R}^3 \times \mathbb{R}^3} dp\, dq\, \widehat{V}(p - k')\widehat{G}^+(|k|, p, q)\widehat{V}(k - q),$$

where $k, k' \in \mathbb{R}^3$, $k = |k|\hat{k}$, $k' = |k|\hat{k}'$,

$$\widehat{V}(p) := \int_{\mathbb{R}^3} dx\, V(x)e^{ip \cdot x},$$

and

$$\widehat{G}^+(|k|, p, q) := \frac{1}{(2\pi)^3} \int_{\mathbb{R}^3 \times \mathbb{R}^3} dx\, dy\, e^{i(q \cdot y - p \cdot x)} G^+(|k|, x, y).$$

Here G^+ is the complete Green's function, i.e., the resolvent kernel $(k^2 + \Delta - V)^{-1}$. The special case of $k' = -k$ becomes

$$\widehat{V}(2k) = -4\pi A(|k|, -\hat{k}, \hat{k})$$
$$- \frac{1}{(2\pi)^3} \int_{\mathbb{R}^3 \times \mathbb{R}^3} dp\, dq\, \widehat{V}(p + k)\widehat{G}^+(|k|, p, q)\widehat{V}(k - q). \qquad (2.110)$$

The resolvent equation for the Green's function reads in the Fourier transform language

$$\widehat{G}^+(|k|, p, q) = \frac{\delta(p-q)}{k^2 - p^2 + i\epsilon}$$

$$+ \frac{1}{(2\pi)^3} \int_{\mathbb{R}^3} dl \, \frac{\widehat{V}(p+l)}{k^2 - p^2 + i\epsilon} \widehat{G}^+(|k|, l, q). \tag{2.111}$$

The idea now is to consider equations (2.110) and (2.111) as a nonlinear system for the Fourier transform $\widehat{V}$ of the potential, in which the *backward* scattering amplitude $A(|k|, -\hat{k}, \hat{k})$ is given. If the scattering amplitude is small enough in a suitable norm, then the mapping defined by (2.111) and the right-hand side of (2.110) may be a contraction and thus it may have a unique fixed point. A detailed study of this system can be found in [Pr69–82]. A more recent investigation by R. G. Novikov and G. M. Henkin [NH87b] encompasses this system. They prove that if the potential has a sufficiently small norm (different from Prosser's) on a specified Sobolev space then it is uniquely determined by the backward scattering amplitude or, alternatively, by other restrictions of the scattering amplitude to three-dimensional manifolds.

It should be noted that this method avoids the large redundancy of the data contained in the function $A(|k|, \theta, \theta')$. The backward scattering amplitude is a function of three variables, as is the potential $V(x)$. Further investigation of this procedure would be desirable.

The Saitō Method. Y. Saitō proved a variant of the large-k result (1.85) [Sa82a].

Theorem 2.9.1. (Saitō) *Suppose that* $\exists C < \infty$, $\mu > 1$ *such that for all* $x \in \mathbb{R}^3$, $|V(x)| < C(1 + |x|)^{-\mu}$. *Then*

$$\lim_{k\to\infty} k^2 \int_{S^2 \times S^2} d\theta \, d\theta' \, A(k, \theta, \theta') e^{ik(\theta - \theta') \cdot x} = -2\pi \int_{\mathbb{R}^3} dy \, \frac{V(y)}{|x - y|^2}. \tag{2.112}$$

His inversion procedure consists of recovering $V(x)$ from a knowledge of the right-hand side of (2.112). For details we refer to his papers [Sa82a–86].

2.10 Notes

2.1 Lemma 2.1.1 is due to Faddeev in [Fa56].

2.2 The method used here is that of [Ne79].

2.3, 2.4 The content of these sections mostly follows [Ne80, 81, 82a, 82b, 83, 85b, 88a]. However, Theorems 2.3.11, 2.4.7, and the use of Lemma 2.4.1 are new.

2.5 The results and methods of this section come from [Ne85b, 88b].

2.6 The definition of the Jost function and the content of this section come from [Ne80, 81, 82a, 82b, 83]. The reduction method of Subsection 2.6.1 was the procedure used in [Ne80, 81, 82a, 82b, and 83]. Theorem 2.6.15 is new. What in Subsection 2.6.2 is called an "operator Riemann-Hilbert problem" differs from the customary statement of a Wiener-Hopf factorization problem. For general discussions of such problems see [GK86 and GMP87]. It should be noted, however, that the assumptions made in the mathematical literature are different from those satisfied by the S matrix. Subsection 2.6.3 follows [Ne82a]. The result of Novikov and Henkin is stated as Theorem 1.5 and Theorem 4.5 of [NH87b].

Note added in proof: The statement of Lemma 2.6.7 is too strong. There is a restriction on the spaces $\mathcal{H}_m$ in $W_\mu^1(S)$, which arises from the fact that the range of a projection with a given nullspace cannot be chosen arbitrarily: It cannot be a subspace of its nullspace. The correct version of the lemma is as follows:

Lemma 2.6.7. *Suppose that the problem $W_\sigma^1(S)$ has a unique solution. Then for every choice of pole positions in another set μ with $n_\mu = n_\sigma$ there exists a choice of spaces $\mathcal{H}_m$ in μ such that $W_\mu^1(S)$ has a holomorphically invertible solution.*

A more detailed version of the proof will be published in the near future.

2.7 This section follows [Ne88a].

2.8 The points of this section were made in [Ne79 and 84].

2.9 The Moses-Prosser method originated with [Mo56] and was continued by [Pr69-82]. There is some recent work on the inverse backscattering problem by Eskin and Ralston [ER88]. The Saitō procedure was the subject of the papers [Sa82a-86].

For a generalization to higher dimensions, see [We89].

Part II

Use of the Regular and Standing-Wave Solutions

3. The Regular Solution

3.1 How to Define a Regular Solution

In the one-dimensional case it is customary and very useful to define a solution of the Schrödinger equation by a k-independent boundary condition at a point. Such a solution is necessarily an entire analytic function of k of exponential type $|x|$. It is this property that is exploited in the solution technique of the inverse spectral problem due to Gel'fand and Levitan.

In higher dimensions appropriate boundary conditions that would ensure that the solution is an entire analytic function of k are not known. A generalization of the Gel'fand-Levitan technique therefore requires that such a solution be defined by another method. It will be done by means of the Jost function. In one dimension the Jost function is defined in terms of the relation between the scattering solution and the regular solution, and its properties are derived from those of these two functions. In higher dimensions this procedure will be inverted.

We shall assume from now on that the operator Riemann-Hilbert problem $W_\sigma^1(S)$ has an invertible solution; this inverse is the Jost function $J(k)$. Section 2.6 is devoted to the details of this question of a Wiener-Hopf factorization of the S matrix. The relevant properties of the Jost function are (see Lemma 2.6.4):

(i) $J(k)$ is a function on $\mathbb{R}$ with values that are operators, $L^2(\mathbb{S}^2) \mapsto L^2(\mathbb{S}^2)$, such that $J - 1$ is in the space $\mathscr{L}^2$ defined in Definition 1.5.12. The function J satisfies the equation

$$J^* = QJQS, \tag{3.1}$$

where S is the S matrix for the given potential, and $J^*(k) := J(-k)$.

(ii) $J(k)$ is the boundary value of an analytic function that is holomorphic in $\mathbb{C}^+$ with zeros at the points $k = i\kappa_m$ if the bound states have the eigenvalues $-\kappa_m^2$, such that $\mathrm{nul} J(i\kappa_m) = \mathscr{H}_m$, where $\mathscr{H}_m$ is the characterspace at the mth eigenvalue; its inverse is meromorphic in $\mathbb{C}^+$ and solves $W_\sigma^1(S)$; furthermore, in $\mathbb{C}^+$

$$\lim_{|k| \to \infty} \|J(k) - 1\| = 0.$$

The *regular solution* of the Schrödinger equation is defined in terms of the scattering solution ψ^+ by

$$\phi(k, \theta, x) := \int_{\mathbb{S}^2} d\theta' \, J(k, \theta, \theta') \psi^+(k, \theta', x). \tag{3.2}$$

Regarding $\psi^+(k, \cdot, x)$ and $\phi(k, \cdot, x)$ as vectors in $L^2(\mathbb{S}^2)$, we may write the relation between ϕ and ψ^+ more simply as

$$\phi(k, x) = J(k)\psi^+(k, x).$$

Since $J(k)$ is independent of x, the function ϕ is, of course, a solution of the Schrödinger equation.

3.2 Properties of the Regular Solution

It follows from (3.1) that the regular solution ϕ satisfies the symmetry

$$\phi(-k, x) = Q\phi(k, x). \tag{3.3}$$

Furthermore, since ψ^+ and J are analytic functions of k such that in $\mathbb{C}^+$ ψ^+ is meromorphic with simple poles the ranges of whose residues equal the nullspaces of J at the same points, ϕ is holomorphic in $\mathbb{C}^+$. The symmetry relation (3.3) then implies that it has an analytic continuation from $\mathbb{C}^+$ into the lower half of the complex plane with no singularities there either. It is therefore an entire analytic function.

Since on the real axis $J(k) - \mathbb{1}$ is in $\mathscr{L}^2$, and since ψ^+ satisfies Lemma 1.4.2 and is uniformly bounded, the function

$$\Phi(k, x) := \phi(k, x) - \phi_0(k, x), \tag{3.4}$$

where $\phi_0(k, x)$ is the $L^2(S^2)$-vector defined by $e^{ik\theta \cdot x}$, is square integrable as a function of k. It also follows from Lemma 1.4.2 that for each fixed $x \in \mathbb{R}^3$ the function $\Phi(k, x)e^{ik|x|}$ is bounded in $\mathbb{C}^+$ and the function $\Phi(k, x)e^{-ik|x|}$ is bounded in $\mathbb{C}^-$. Therefore we may conclude the following.

Lemma 3.2.1. *For each $x \in \mathbb{R}^3$ the function Φ defined by (3.4) as a function of k has a Fourier transform in the norm-L^2 sense and the support of that Fourier transform lies in the interval $[-|x|, |x|]$ such that*

$$\Phi(k, x) = \int_{-|x|}^{|x|} dt \, e^{ikt} \widehat{\Phi}(t, x).$$

This lemma establishes the sense in which the regular solution has the generalized **Povzner-Levitan representation**

$$\phi(k, \theta, x) = e^{ik\theta \cdot x} - \int_{-|x|}^{|x|} dt \, e^{ikt} w(t, \theta, x). \tag{3.5}$$

The symmetry (3.3) implies the following symmetry for the Fourier transform function w

$$w(-t, -\theta, x) = w(t, \theta, x). \tag{3.6}$$

Thus the solution of the Schrödinger equation defined by (3.2) is an entire analytic function of k, of exponential type $|x|$, and is an appropriate generalization of the regular solution defined by a boundary condition at a point in the one-dimensional case. Its existence, however, depends on the existence of a solution to the Wiener-Hopf factorization of the S matrix that leads to the Jost function.

Let us insert (3.5) in the Schrödinger equation. This leads to the hyperbolic partial differential equation for w

$$\left(\varDelta - \frac{\partial^2}{\partial t^2}\right) w = Vw \tag{3.7}$$

in the regions $-|x| < t < \theta \cdot x$, $\theta \cdot x < t < |x|$; for $t = \pm|x|$ it must vanish,

$$w(t = \pm|x|, \theta, x) = 0, \tag{3.8}$$

and on the plane $t = \theta \cdot x$ its normal derivative must have a discontinuity

$$\theta \cdot \nabla[w(t = \theta \cdot x+, \theta, x) - w(t = \theta \cdot x-, \theta, x)] = \tfrac{1}{2}V(x). \tag{3.9}$$

Of course, it has to be remembered that the function w is not known to be differentiable with respect to x. Thus the sense in which these equations hold is not completely clear.

These differential equations can be somewhat simplified by the use of the symmetry (3.6). The function w can be formed out of a function u as

$$w(t, \theta, x) = \begin{cases} u(t, \theta, x), & \text{if } -|x| < t < \theta \cdot x, \\ u(-t, -\theta, x), & \text{if } \theta \cdot x < t < |x|, \end{cases}$$

and u solves the quasi-Goursat problem in the region $-|x| < t < \theta \cdot x$

$$\left(\varDelta - \frac{\partial^2}{\partial t^2}\right) u = Vu, \quad u(-|x|, \theta, x) = 0, \quad \theta \cdot \nabla u(\theta \cdot x, \theta, x) = -\tfrac{1}{4}V(x). \tag{3.10}$$

The boundaries here are the cone $|x| = -t$ and its tangent plane normal to θ. Nothing seems to be known about conditions under which this problem has a solution. If, for a certain class of potentials, the existence of a solution could be proved, that could serve to prove the existence of the regular solution of the Schrödinger equation and it would obviate the need to define ϕ via the Jost function. There cannot, however, be more than one solution of this system of equations.

Lemma 3.2.2. *The system (3.6)–(3.9) has at most one solution.*

Proof. We prove this by examining the system (3.10). The proof then proceeds like that of Lemma 2.4.1. The only difference is that the system in $\mathbb{R}^2$ is exterior to the circle $C = \{x, y \mid x^2 + y^2 = -4\xi\zeta\}$, with the boundary condition that the solutions vanish on C. $\qquad\square$

Note that Lemma (3.2.2) implies that, since the potential is real, the function w is real.

While the Fourier transform of the regular solution with respect to the variable k is given by (3.5), it will be more useful for later purposes to define a three-dimensional Fourier transform with respect to the variable $k\theta$

$$h(x, y) := \frac{1}{(2\pi)^3} \int_0^\infty dk\, k^2 \int_{s^2} d\theta\, e^{-ik\theta \cdot y} [e^{ik\theta \cdot x} - \phi(k, \theta, x)]. \tag{3.11}$$

The extra factor of k^2 in the integrand makes the integral in general neither absolutely convergent nor convergent in the mean, and it will have to be taken in the sense of distributions.

If the Fourier transform (3.11) is inserted in the Schrödinger equation one obtains the following hyperbolic partial differential equation that has to be satisfied by $h(x, y)$ in the region $|y| \leq |x|$

$$(\Delta_x - \Delta_y)h(x, y) = V(x)h(x, y). \tag{3.12}$$

Furthermore, h has to satisfy the boundary condition on $|x| = |y|$

$$-2\hat{x} \cdot \nabla_x[h(|x|\hat{x}, |x|\hat{y})|x|^2] = \delta(\hat{x}, \hat{y})V(x). \tag{3.13}$$

Here Δ_x and Δ_y stand for the Laplacians with respect to x and y, respectively, ∇_x is the gradient with respect to x, and $\hat{x} = x/|x|$. These equations and boundary conditions for h are, of course, the analogues of (3.7) and (3.9) for w.

The three-dimensional and one-dimensional Fourier transforms of ϕ are related to one another by the Radon transform:

$$w(t, \theta, x) = \int_{\mathbb{R}^3} dy\, h(x, y)\delta(t - \theta \cdot y). \tag{3.14}$$

The question now is whether the fact that the support of w as a function of t lies in the interval $[-|x|, |x|]$ allows us to conclude that the support of h as a function of y lies in the ball $|y| \leq |x|$. A sufficient condition, applicable to distributions, for this to be the case is that w satisfy the symmetry (3.6) and that ϕ satisfy the equations

$$\int_{S^2} d\theta\, \phi^{(n)}(0, \theta, x)Y_L(\theta) = 0 \tag{3.15}$$

for all $l > n$, where $\phi^{(n)}$ denotes the nth derivative of ϕ with respect to k and the functions Y_L are spherical harmonics in the notation introduced in Section 1.6. Under these conditions w is, in fact, a Radon transform of some function h and the intuitively expected support relation between w and h holds. The following result will be proved in Section 3.2.1.

Lemma 3.2.3. *If $\dot{V} \in \mathcal{W}$ and it satisfies (1.80) for some $\epsilon > 0$ then in general the regular solution ϕ satisfies (3.15) and thus is a Radon transform.*

Remark. For central potentials this behavior of the regular solution is well known and its proof does not require the strong assumption of exponential decay of the potential. That is because for central potentials the regular solution can be defined directly as a solution of a Volterra integral equation, whereas in the noncentral case its definition requires the circuitous route via the Wiener-Hopf factorization of the S matrix and the Jost function.

The spherical harmonics have the symmetry $Y_L(-\theta) = (-1)^l Y_L(\theta)$; therefore the definition of $\phi_L(k, x)$ in Section 3.4 shows that it may be expressed in the form

$$\phi_L(k, x) = k^l \widehat{\phi}(k^2, x),$$

where $\widehat{\phi}$ is an entire analytic function of k^2. Now one may choose such linear combinations of spherical harmonics to form the functions we call Y_L here that $k^l Y_L(\theta)$ is a homogeneous polynomial of degree l in the three Cartesian components of the vector $k\theta = (k_1, k_2, k_3)$. It then follows from (3.30) and the analyticity of ϕ as a function of k that $\phi(k, \theta, x)$ is an entire analytic function of each of the Cartesian components of $k\theta$ separately.

Let us summarize these results.

Theorem 3.2.4. *Suppose that $V \in \mathcal{W}$. If the Jost function exists as an invertible solution of the Wiener-Hopf problem $W_\sigma^1(S)$ then there exists a regular solution $\phi(k, \theta, x)$ of the Schrödinger equation in $\mathbb{R}^3$ given by (3.2) with the following properties:*

(i) ϕ satisfies the symmetry (3.3);

(ii) as a function of k, $\phi(k, \theta, x) - e^{ik\theta \cdot x}$ is $L^2(\mathbf{S}^2)$-norm square integrable in $\mathbb{R}$ for each $x \in \mathbb{R}^3$;

(iii) for each $x \in \mathbb{R}^3$ and almost all $\theta \in \mathbf{S}^2$, $\phi(k, \theta, x)$ is an entire analytic function of k of exponential type $|x|$.

If, furthermore, V satisfies (1.80) for some $\epsilon > 0$, then in general

(iv) each expansion coefficient of ϕ as a function of θ on the basis of the spherical harmonics satisfies (3.30), or, equivalently, ϕ satisfies (3.15) (exceptions have not been ruled out);

(v) ϕ is an entire analytic function of each Cartesian component of $k\theta$ separately.

Assuming that all the conditions are satisfied for ϕ to have the needed properties, its three-dimensional Fourier transform, which may exist only in the sense of distributions, has its support entirely in the ball $\mathcal{B}(|x|)$ of radius $|x|$,

$$\mathcal{B}(r) := \{y \in \mathbb{R}^3 \mid |y| < r\},$$

and we have another generalized Povzner-Levitan representation:

$$\phi(k, \theta, x) = \phi_0(k, \theta, x) - \int_{\mathcal{B}(|x|)} dy\, h(x, y)\phi_0(k, \theta, y), \qquad (3.16)$$

where $\phi_0(k, \theta, x) = e^{ik\theta \cdot x}$. We also note that the inverse of the Radon transform (3.14) is given by

$$h(x, y) = -\frac{1}{8\pi^2}\Delta_y \int_{\mathbf{S}^2} d\theta\, w(\theta \cdot y, \theta, x). \qquad (3.17)$$

The function h is *real*.

It will simplify matters considerably if equation (3.16) is written in an operator notation,

$$\phi = U\phi_0.$$

(Notice that this is not the same notation we used earlier when the dependence of functions on θ was operated on. In the present case it is an integration over $\mathbb{R}^3$ that is suppressed in the notation.) The operator U is given by $U = \delta - h$ or the kernel

$$U(x, y) = \delta(x - y) - h(x, y),$$

and $h(x, y) = 0$ for $|y| > |x|$. Such a kernel is usually referred to as "lower triangular".

Let us construct the inverse of U. The kernel of the right inverse of U, $(U^{-1})_r = \delta - h'$ must satisfy the Volterra equation

$$h'(x, y) = -h(x, y) + \int_{\mathscr{B}(|x|)} dz\, h(x, z)h'(z, y), \tag{3.18}$$

which has a unique solution and this solution $h'(x, y)$ vanishes when $|x| < |y|$. The kernel of the left inverse of U, $(U^{-1})_l = \delta - h''$ must satisfy the equation

$$h''(x, y) = -h(x, y) + \int_{\mathscr{B}'(|y|)} dz\, h''(x, z)h(z, y), \tag{3.19}$$

where $\mathscr{B}'(r)$ is the exterior of $\mathscr{B}(r)$.

The equation $Uf = 0$ implies that f satisfies the homogeneous version of (3.18) and hence $f = 0$. Therefore the left inverse of U exists and (3.19) has a solution. Furthermore, this solution of (3.19) must be unique, because otherwise its homogeneous form would have a nontrivial solution, which would imply that there must be an f such that $\tilde{U}f = 0$. But then a right inverse could not exist. Now for $|y| > |x|$ equation (3.19) becomes

$$h''(x, y) = \int_{\mathscr{B}'(|y|)} dz\, h''(x, z)h(z, y).$$

If this equation had a nontrivial solution then the solution of (3.19) would not be unique. Consequently the unique solution of (3.19) must be lower triangular, i.e., $h''(x, y) = 0$ for $|y| > |x|$. Since left and right inverses of U exist and are unique, they are equal.

Lemma 3.2.5. The operator $U = \delta - h$ of the Povzner-Levitan representation is invertible and its inverse is lower triangular: $(U^{-1}) = \delta - h'$, $h'(x, y) = 0$ for $|y| > |x|$.

The inverse of (3.16) reads, explicitly,

$$\phi_0(k, \theta, x) = \phi(k, \theta, x) + \int_{\mathscr{B}(|x|)} dy\, l(x, y)\phi(k, \theta, y). \tag{3.20}$$

The representations (3.16) and (3.20) may be analytically continued, as functions of k, into any part of the complex plane; because of the lower triangularity of the kernels the integrals always converge. In particular, they hold also for the bound states. The corresponding function ϕ_0 is then, of course, not in $L^2(\mathbb{R}^3)$, whereas ϕ is. We must examine $\phi(i\kappa, \theta, x)$.

The Jost function was defined so that at a bound-state (simple) pole of ψ^+ its nullspace equals the range of ψ's residue, which is given by (1.31). This implies that if $-\kappa^2$ is an m-fold degenerate eigenvalue and Y_κ^b are a set of character functions then

$$J(i\kappa)QY_\kappa^b = 0.$$

We denote the k-derivative of J by J'. Then it follows from (1.31) that

$$\phi(i\kappa, x) = \frac{1}{2i\kappa}J'(i\kappa)Q\sum_{a,b} Y_\kappa^a d_{ab}^\kappa u_\kappa^b(x) \tag{3.21}$$

in vector notation. Let us choose m real functions $X_\kappa^b(\theta)$ that form a bi-orthogonal set with the real functions $J'(i\kappa)QY_\kappa^b/2i\kappa$ so that

$$\left(X_\kappa^a, J'(i\kappa)QY_\kappa^b\right) = 2i\kappa\delta_{ab}. \tag{3.22}$$

Then

$$\left(X_\kappa^a, \phi(i\kappa, x)\right) = \sum_b d_{ab}^\kappa u_\kappa^b(x) \tag{3.23}$$

$$= \int_{\mathbb{R}^3} dy\, U(x, y)\left(X_\kappa^a, \phi_0(i\kappa, y)\right)$$

by (3.16). If we define

$$Z_\kappa^a(x) := \sum_b e_{ab}^\kappa \int_{\mathbb{S}^2} d\theta\, X_\kappa^b(\theta) e^{-\kappa\theta\cdot x}, \tag{3.24}$$

then this means that

$$u_\kappa^a(x) = \int_{\mathbb{R}^3} dy\, U(x, y) Z_\kappa^a(y). \tag{3.25}$$

Note that, although the functions $Z_\kappa^a(x)$ are exponentially increasing, the integral in (3.25) not only converges for each x, but as $|x| \to \infty$ it vanishes exponentially.

Thus the operator U whose kernel is $U(x, y)$ will be regarded as $\mathscr{H}' \mapsto \mathscr{H}$, and its inverse, $\mathscr{H} \mapsto \mathscr{H}'$. Here $\mathscr{H} = L^2(\mathbb{R}^3)$, and $\mathscr{H}'$ is $\mathscr{H}$ enlarged by the addition of the finite number of exponentially increasing functions $Z_{\kappa_n}^a(x)$.

3.2.1 Proof of Lemma 3.2.3

We approach the question whether the regular solution satisfies the needed conditions (3.15) by means of its angular-momentum projections. Let us first define the partial projection of the function $\mathscr{L}(k) = J(k) - \mathbb{1}$ defined at the beginning of Section 2.6.3:

$$\mathscr{L}_L(k, \theta) := \int_{\mathbb{S}^2} d\theta'\, \overline{Y}_L(\theta')\mathscr{L}(k, \theta, \theta').$$

Now,

$$\int_{\mathbb{S}^2} d\theta\, \mathscr{L}_L(k, \theta) e^{ik\theta\cdot x} = 4\pi \sum_{L' \leq L} \mathscr{L}_{L'L}(k) i^{l'} \overline{Y}_{L'}(\hat{x}) j_{l'}(kr)$$

$$+ \int_{\mathbb{S}^2} d\theta\, \mathscr{L}_L(k, \theta) E_L(k, \theta, x), \tag{3.26}$$

where

$$E_L(k, \theta, x) := e^{ik\theta\cdot x} - 4\pi \sum_{L' \leq L} i^{l'} Y_{L'}(\theta)\overline{Y}_{L'}(\hat{x}) j_{l'}(kr),$$

$\hat{x} = x/|x|$, $r = |x|$, j_l is the spherical Bessel function and $L' \leq L$ means that $l' \leq l$. It follows from Lemma 2.6.16 that the sum on the right-hand side of (3.26) is $O(k^l)$ as $k \to 0$, with the leading term in k being $O(r^l)$ as $r \to \infty$. The function E_L, on the other hand is $O(k^{l+1})$ with the leading term being $O(r^{l+1})$. Since $\widetilde{J}(k)$ is assumed to be a solution of $W^1_\sigma(S^*)$, it is in $\mathscr{L}^2$ (see Definition 1.5.12) and hence

$$\int_{-\infty}^{\infty} dk \int_{S^2} d\theta \, |\mathscr{L}_L(k,\theta)|^2 < \infty,$$

which implies that

$$\lim_{k\to 0} k^2 \int_{S^2} d\theta \, |\mathscr{L}_L(k,\theta)|^2 = 0.$$

As a result, the integral in (3.26) is $O(k^l)$, with the leading term being $O(r^{l+1})$, and we have as a result

$$\int_{S^2} d\theta \, \mathscr{L}_L(k,\theta) e^{ik\theta\cdot x} = O(k^l) \tag{3.27}$$

as $k \to 0$, with the leading term being $O(r^{l+1})$ as $r \to \infty$.

We now set up an integral equation for the regular solution. The definition (3.2) of ϕ together with the Lippmann-Schwinger equation (1.3) directly leads to the integral equation

$$
\begin{aligned}
\phi(k,\theta,x) \;=\; & e^{ik\theta\cdot x} + \int_{S^2} d\theta' \, \mathscr{L}(k,\theta',\theta) e^{ik\theta'\cdot x} \\
& - \int_{\mathbb{R}^3} dy \, \frac{e^{ik|x-y|}}{4\pi|x-y|} V(y)\phi(k,\theta,y)
\end{aligned}
\tag{3.28}
$$

for ϕ. Fourier transformation of this equation with respect to the variable k leads to the following equation for the function w defined by (3.5), for $-|x| < t < |x|$

$$
\begin{aligned}
w(t,\theta,x) \;=\; & \xi(t,\theta,x) - \int_{S^2} d\theta' \, \mu(t - \theta'\cdot x,\theta,\theta') \\
& - \int_{\Omega(t,x)} dy \, \frac{V(y)}{4\pi|x-y|} w(t - |x-y|,\theta,y),
\end{aligned}
\tag{3.29}
$$

where the region of integration $\Omega(t,x)$ is given by

$$\Omega(t,x) = \{ y \in \mathbb{R}^3 \mid -|y| < t - |x-y| < |y| \},$$

and

$$\mu(t,\theta,\theta') := \frac{1}{2\pi} \int_{-\infty}^{\infty} dk \, e^{-ikt} [J(k,\theta,\theta') - \delta(\theta,\theta')],$$

$$\xi(t,\theta,x) := \int_{\mathbb{R}^3} dy \, \frac{V(y)}{4\pi|x-y|} \delta(|x-y| + \theta\cdot y - t).$$

The angular-momentum projection of (3.28) leads to an integral equation for the function

$$\phi_L(k,x) := \int_{S^2} d\theta \, \overline{Y_L(\theta)} \phi(k,\theta,x),$$

namely,

$$\phi_L(k,\theta) = 4\pi i^l j_l(kr)\overline{Y_L(\hat{x})} + \int_{S^2} d\theta \, \mathcal{L}_L(k,\theta)e^{ik\theta\cdot x}$$

$$-\frac{1}{4\pi}\int_{\mathbb{R}^3} dy \, \frac{e^{ik|x-y|}}{|x-y|}V(y)\phi_L(k,y).$$

By (3.27) and the behavior of the spherical Bessel function near the origin, the inhomogeneity in this integral equation is $O(k^l)$ as $k \to 0$, with the leading term being $O(r^{l+1})$ as $r \to \infty$. It then easily follows that if $k = 0$ is not exceptional and if $V \in \mathcal{W}$ satisfies (1.80) for some $\epsilon > 0$ then

$$\phi_L(k,x) = O(k^l) \tag{3.30}$$

as $k \to 0$. Since ϕ is an entire analytic function of k, this result implies (3.15) *in general*. It may not hold in exceptional instances whose occurrence I have not been able to rule out.   □

3.3 Completeness

Section 1.3 contained a discussion of the completeness of the scattering solutions of the Schrödinger equation. We shall write the statement (1.45) of this completeness in the following way:

$$\frac{1}{(2\pi)^3}\int_0^\infty dk\, k^2\Big(\psi^+(k,x),\psi^+(k,y)\Big) + \sum_{n,a,b} u^a_{\kappa_n}(x)d^{\kappa_n}_{ab}u^b_{\kappa_n}(y) = \delta(x-y), \tag{3.31}$$

where the inner product is on $L^2(S^2)$ and the bound-state eigenfunctions are normalized as in (1.27), with the d^κ_{ab} defined there. It follows from equation (3.23) that

$$\sum_{a,b} u^a_\kappa(x)d^\kappa_{ab}u^b_\kappa(y) = \Big(\phi(i\kappa,x), M_\kappa\phi(i\kappa,y)\Big), \tag{3.32}$$

where

$$M_\kappa(\theta,\theta') := \sum_{a,b} X^a_\kappa(\theta)e^\kappa_{ab}X^b_\kappa(\theta'). \tag{3.33}$$

By Lemma 1.5.7 a set of character functions $Y^a_\kappa(\theta)$, $a = 1,\ldots,N$ can be determined from the scattering amplitude. According to Corollary 1.2.6 one can then use equations (1.41) and (1.66) to determine the corresponding matrix d_κ, whose inverse is the matrix e_κ with entries e^κ_{ab}. Therefore we have the following.

Lemma 3.3.1. *If the eigenvalue* $-\kappa^2$ *is normal (see Definition 1.2.5) then the kernel* M_κ *can be determined from the scattering amplitude.*

The matrix e_κ is the analogue of the *norming constants* in one dimension and for the radial Schrödinger equation.

We are now ready to translate (3.31) into its equivalent equation for ϕ. Inserting the relation (3.2) between the regular solution and the scattering solution in (3.31), one obtains the following form of the completeness relation for the regular solution

$$\int \left(\phi(k, x), d\rho(k)\phi(k, y) \right) = \delta(x - y), \tag{3.34}$$

where the *spectral function* is given by

$$\frac{d\rho}{dE} = \begin{cases} \frac{k}{16\pi^3}[M(k) + \mathbb{1}], & \text{if } E > 0, \\ \sum_n M_{\kappa_n}\delta(E - E_n), & \text{if } E < 0. \end{cases} \tag{3.35}$$

Here

$$M(k) := F(k)^\dagger F(k) - \mathbb{1},$$

where $F(k)$ is the inverse of the Jost function, $F = J^{-1}$, and $E := k^2$. The integral in (3.34) is meant to run over the entire range $-\infty < E < \infty$. When $V = 0$ this equation goes over into the Fourier theorem

$$\int \left(\phi_0(k, x), d\rho_0(k)\phi_0(k, y) \right) = \delta(x - y),$$

and

$$\frac{d\rho_0}{dE} = \begin{cases} \frac{k}{16\pi^3}\mathbb{1}, & \text{if } E > 0, \\ 0, & \text{if } E < 0. \end{cases}$$

The spectral function is obviously a positive semi-definite operator.

Lemma 3.3.2. *For every $k \in \mathbb{R}$ for which the Jost function $J(k)$ exists the spectral function is positive definite.*

Proof. If $(M + \mathbb{1})f = 0$ then necessarily $Ff = 0$ and F cannot have an inverse; thus J fails to exist. □

3.4 Notes

3.1 The definition of the regular solution by means of the Jost function originated in [Ne80].

3.2 In [Ne80] the properties of the regular solution were stated incorrectly; this was rectified in [Ne81], and the correct Povzner-Levitan representation was given. The quasi-Goursat problem (3.10) was discussed in [Ne87]. An appropriate reference for the Radon transform is [Lu66]. The procedure of using angular-momentum projections to prove Lemma 3.2.3 and Theorem 3.2.4 was used in [Ne82a].

3.3 This section is based on [Ne80].

4. The Inverse Problem

4.1 Generalized Gel'fand-Levitan Equations

All the ingredients for the solution of the inverse scattering problem via the Gel'fand-Levitan procedure have now been assembled. What is actually going to be solved is an inverse spectral problem posed for the regular solution. This inverse spectral problem is of relatively little intrinsic interest because in dimensions higher than one (for noncentral potentials) the regular solution is not a natural solution of the Schrödinger equation. As we saw in Chapter 3, it has to be defined in a very indirect manner and the inverse spectral problem for it does not arise naturally. However, solving this inverse spectral problem, in which the spectral function, defined in Section 3.3, is the input, solves at the same time the inverse scattering problem because the Jost function forms a direct link from the scattering data, that is, the S matrix, to the spectral function. Thus, solving the problem of finding the potential that underlies a given spectral function also solves the problem of finding the potential that underlies a given S matrix. Once the Wiener-Hopf factorization problem has been solved and the Jost function has been constructed from the S matrix, the spectral function is known.

It will be useful to employ the symbolism introduced near the end of Section 3.2, in which U is the Povzner-Levitan operator that converts the regular solution for $V = 0$ into that for V. Let us indicate that explicitly by calling it U_{V0} and the regular solution for V, ϕ_V. Lemma 3.2.5 states that this operator has an inverse:

$$\phi_0 = \mathsf{U}_{V0}^{-1}\phi_V = \mathsf{U}_{0V}\phi_V,$$

where the kernels are $U_{V0} = \delta - h_{V0}$, $U_{0V} = \delta - h_{0V}$ and both $h_{V0}(x, y)$ and $h_{0V}(x, y)$ are lower triangular, i.e., they vanish when $|y| > |x|$. We then have

$$\phi_W = \mathsf{U}_{W0}\phi_0 = \mathsf{U}_{W0}\mathsf{U}_{0V}\phi_V = \mathsf{U}_{WV}\phi_V,$$

where the kernel $U_{WV} = \delta - h_{WV}$ and

$$h_{WV}(x, y) = h_{W0}(x, y) + h_{0V}(x, y) - \int_{\mathscr{B}(|x|)} dz\, h_{W0}(x, z)h_{0V}(z, y).$$

It follows that $h_{WV}(x, y)$ vanishes for $|y| > |x|$. We also have $\mathsf{U}_{WV} = \mathsf{U}_{VW}^{-1}$. The kernel U_{VW} connects the regular solutions of the Schrödinger equations with V and W:

$$\phi_V(k, \theta, x) = \phi_W(k, \theta, x) - \int_{\mathscr{B}(|x|)} dy\, h_{VW}(x, y)\phi_W(k, \theta, y). \tag{4.1}$$

The completeness relation (3.34) may be written symbolically

$$\int (\phi_V, d\rho_V \phi_V) = \delta.$$

Since $\phi_V = U_{VW}\phi_W$, $U_{VW}^{-1} = U_{WV}$, and U_{VW} is real this leads to

$$U_{WV} = \int (\phi_W, d\rho_V \phi_W)\tilde{U}_{VW}$$

and

$$\tilde{U}_{VW} = \int (\phi_W, d\rho_W \phi_W)\tilde{U}_{VW},$$

where the tilde denotes the operator with the transposed kernel. Subtraction yields for the kernels

$$U_{WV} - \tilde{U}_{VW} = \tilde{h}_{VW} - h_{WV} = \int \left(\phi_W, d(\rho_V - \rho_W)\phi_W\right)\tilde{U}_{VW}.$$

Similarly one obtains

$$U_{WV}\tilde{U}_{WV} - \delta = h_{WV}\tilde{h}_{WV} - h_{WV} - \tilde{h}_{WV} = \int \left(\phi_W, d(\rho_V - \rho_W)\phi_W\right).$$

Writing more explicitly

$$r_{VW}(x, y) := \int \left(\phi_W(k, x), d[\rho_V(k) - \rho_W(k)]\phi_W(k, y)\right), \tag{4.2}$$

and using the lower triangularity of h_{VW} and h_{WV}, one finds that the first of these equations reads for $|x| > |y|$

$$h_{VW}(x, y) = r_{VW}(x, y) - \int_{\mathscr{B}(|x|)} dz\, r_{VW}(y, z)h_{VW}(x, z), \tag{4.3}$$

and the second reads for $|x| > |y|$

$$h_{WV}(x, y) = -r_{VW}(x, y) + \int_{\mathscr{B}(|y|)} dz\, h_{WV}(y, z)h_{WV}(x, z). \tag{4.4}$$

The first is the *generalized* linear *Gel'fand-Levitan equation*, and the second is the *generalized* nonlinear *Gel'fand-Levitan equation*. The function that serves as their input, $r_{VW}(x, y)$, is real and symmetric: $r_{VW}(x, y) = r_{VW}(y, x) = \overline{r_{VW}(x, y)}$.

It follows from (4.2), (4.3), and (4.1) that $h(x, y)$ has the representation

$$h_{VW}(x, y) = \int \left(\phi_W(k, x), d[\rho_V(k) - \rho_W(k)]\phi_V(k, y)\right). \tag{4.5}$$

These are the equations for the functions that lead from the solutions of the Schrödinger equation with a known potential W to the solutions for another potential V. If the "unperturbed" potential is taken to be $W = 0$ then the input function becomes

$$r(x, y) = \int \left(\phi_0(k, x), d[\rho(k) - \rho_0(k)]\phi_0(k, y)\right), \tag{4.6}$$

or, more explicitly, by (3.35) and (3.24)

$$r(x, y) = \frac{1}{(2\pi)^3} \int_0^\infty dk\, k^2 \left(\phi_0(k, x), M(k)\phi_0(k, y) \right)$$
$$+ \sum_{n,a,b} d_{ab}^{\kappa_n} Z_{\kappa_n}^a(x) Z_{\kappa_n}^b(y). \tag{4.7}$$

The generalized Gel'fand-Levitan equation then reads for $|x| > |y|$

$$h(x, y) = r(x, y) - \int_{\mathscr{B}(|x|)} dz\, r(y, z)h(x, z). \tag{4.8}$$

Lemma 4.1.1. *If the Jost function exists for almost all $k \in \mathbb{R}$ then the operator r whose kernel is $r(x, y)$ is such that $r + \mathbb{1}$ is positive definite.*

Proof. Let r^c be the part of r that comes from the continuous spectrum. Then by (3.35)

$$\int_{\mathbb{R}^3 \times \mathbb{R}^3} dx\, dy\, \overline{f}(x)[r^c(x, y) + \delta(x - y)]f(y)$$
$$= \frac{1}{(2\pi)^3} \int_0^\infty dk\, k^2 \|F(k)\widehat{f}(k)\|^2,$$

where $\widehat{f}$ is the Fourier transform of f and $\|\cdot\|$ is the $L^2(\mathbb{S}^2)$-norm. Thus the left-hand side vanishes if and only if $F(k)\widehat{f}(k) = 0$ for almost all $k \in \mathbb{R}$, and hence $J(k)$ fails to exist. The part of r that refers to the point spectrum is semi-definite. Hence $r + \mathbb{1}$ is positive definite.   □

As compared with the generalized Marchenko equation the generalized Gel'fand-Levitan equation has the virtue of being an integral equation on a finite interval. (Note that it is not a quasi-Volterra equation.) However, in order to compute the spectral function, which serves as its kernel and its inhomogeneity, one has to solve first a generalized Marchenko equation for the Jost function. Furthermore, its kernel and inhomogeneity $r(x, y)$ can generally be expected to be a distribution rather than a compact operator, as is the kernel of the generalized Marchenko equation, and one is looking for distributional solutions of (4.8). Lemma 4.1.1 implies that the spectrum of r lies to the right of -1. Thus the solution of the generalized Gel'fand-Levitan equation is unique. However, since r is not known to be compact and we are looking for distributional solutions of (4.8), this does not necessarily ensure the existence of a solution of (4.8).

Lemma 4.1.2. *Suppose that h satisfies the generalized Gel'fand-Levitan equation (4.8). Then h also satisfies the nonlinear equation (4.4) for $W = 0$ and the functions $\phi = U\phi_0$ satisfy the completeness relation (3.34).*

Proof. Assume that $h(x, y)$ solves (4.8) for $|x| > |y|$; set it equal to zero for $|x| < |y|$. Define $h'^\dagger := h - r + hr$ [in the sense of the integration in (4.8)], so that h' is lower triangular. Inserting the definition of h', we have $h' + h - h'h = \alpha$, where $\alpha := \delta - (\delta - h)(\delta + r)(\delta - h^\dagger)$. The left-hand side of this equation is lower triangular, while $\widetilde{\alpha} = \overline{\alpha}$. Therefore the left-hand side vanishes both for $|x| \leq |y|$ and for $|x| \geq |y|$, i.e., $h' + h - h'h = 0$, and hence, $(\delta - h')(\delta - h) = \delta$, which implies that $h'_{V0} = h_{0V}$ and $\delta - h' = U^{-1}$. Thus equation (4.8) becomes equivalent to

$U(\mathbb{1}+r)U^\dagger = \delta$, which is the completeness statement, and $r = U^{-1}U^{\dagger-1} - \delta = -h_{0V} - h^\dagger_{0V} + h_{0V}h^\dagger_{0V}$, which is (4.4).   □

As a next step it is incumbent upon us to check whether the solution of the generalized Gel'fand-Levitan equation leads to a solution of the Schrödinger equation. This is done by using the fact that the function $r(x, y) = r(y, x)$ satisfies the partial differential equation

$$(\Delta_y - \Delta_x)r(x, y) = 0$$

and defining

$$\xi(y, x) := (\Delta_y - \Delta_x)h(y, x).$$

Differentiating equation (4.8) and integrating by parts leads to the following integral equation for $|x| > |y|$:

$$\xi(x, y) = \int_{S^2} d\hat{z}\, r(\hat{z}|x|, y) V(x, |x|\hat{z}) - \int_{\mathscr{B}(|x|)} dz\, \xi(x, z)r(z, y),$$

where

$$V(x, |x|\hat{z}) := -2\hat{x} \cdot \nabla_x[|x|^2 h(x, |x|\hat{z})]. \tag{4.9}$$

If the generalized Gel'fand-Levitan equation (4.8) has a unique solution then it follows that

$$\xi(x, y) = \int_{S^2} d\hat{z}\, V(x, |x|\hat{z})h(|x|\hat{z}, y).$$

Therefore for $|x| > |y|$, $h(x, y)$ satisfies the following partial differential equation:

$$(\Delta_x - \Delta_y)h(x, y) = \int_{S^2} d\hat{z}\, V(x, |x|\hat{z})h(|x|\hat{z}, y). \tag{4.10}$$

Use of this equation in the Povzner-Levitan representation (3.16) and two integrations by parts lead to the Schrödinger equation

$$(\Delta + k^2)\phi(k, \theta, x) = \int_{S^2} d\hat{z}\, V(x, |x|\hat{z})\phi(k, \theta, |x|\hat{z}), \tag{4.11}$$

with a *nonlocal* potential of the type first introduced by Kay and Moses [KM61a]. If the potential (4.9) reduces to the form given by (3.13) then it is local. Thus we have the following important result.

Theorem 4.1.3. *Suppose that a spectral function is given as a positive definite operator-valued function and that $r(x, y)$ is defined in terms of it as in (4.6). Let $h(x, y)$ solve (4.8) for $|x| > |y|$ and let ϕ be given by (3.16) in terms of it. Then $\phi(k, \theta, x)$ satisfies the Schrödinger equation (4.11) with the potential (4.9) and the functions ϕ form a complete set in the sense of (3.34) with the spectral function that was given initially. If (4.9) reduces to the form (3.13), the potential is local.*

It is also possible to obtain an equation for the function w of (3.5). This is done by taking a partial Radon transform of (4.8), which holds for $|x| < |y|$ only. Define

$$s_{|x|}(t, \theta, z) := \int_{\mathscr{B}(|x|)} dy\, r(z, y) \delta(t - \theta \cdot y).$$

Then (4.8) and (3.14) yield, for $t^2 < |x|^2$,

$$w(t, \theta, x) = s_{|x|}(t, \theta, x) - \int_{\mathscr{B}(|x|)} dz\, s_{|x|}(t, \theta, z) h(x, z).$$

Now in view of (4.6) the function $s_{|x|}(t, \theta, z)$ can be expressed in the form

$$s_{|x|}(t, \theta, z) = \int_{S^2} d\theta'\, \mathscr{U}_{|x|}(\theta, \theta', t, \theta' \cdot z),$$

and because of (3.14)

$$\int_{\mathbb{R}^3} dz\, h(x, z) f(\theta \cdot z) = \int_{-\infty}^{\infty} dt\, w(t, \theta, x) f(t).$$

Therefore the partial Radon transform of (4.8) becomes the following generalized Gel'fand-Levitan equation for w, $t^2 < |x|^2$,

$$\begin{aligned} w(t, \theta, x) &= \int_{S^2} d\theta'\, \mathscr{U}_{|x|}(\theta, \theta', t, \theta' \cdot x) \\ &\quad - \int_{S^2} d\theta' \int_{-|x|}^{|x|} dt'\, \mathscr{U}_{|x|}(\theta, \theta', t, t') w(t', \theta', x). \end{aligned} \tag{4.12}$$

The function $\mathscr{U}$ indirectly defined above can be expressed by means of the following formula, valid for $|t| < s$, $s \in \mathbb{R}_+$, $t \in \mathbb{R}$,

$$\begin{aligned} \mathscr{J}(k, \theta \cdot \theta', s, t) &:= \int_{\mathscr{B}(s)} dy\, e^{ik\theta' \cdot y} \delta(t - \theta \cdot y) \\ &= \frac{2\pi \sqrt{s^2 - t^2}}{k|\theta \times \theta'|} e^{ikt\theta \cdot \theta'} J_1\!\left(k\sqrt{s^2 - t^2}|\theta \times \theta'|\right), \end{aligned}$$

where $|\theta \times \theta'| = \sqrt{1 - (\theta \cdot \theta')^2}$ and J_1 is the Bessel function of order one. By (3.35) this leads to the following expression for $\mathscr{U}$

$$\begin{aligned} \mathscr{U}_{|x|}(\theta, \theta', t, t') &= \frac{1}{(2\pi)^3} \int_0^{\infty} dk\, k^2 \int_{S^2} d\theta''\, e^{ikt'} \mathscr{J}(k, \theta \cdot \theta'', |x|, t) M(k, \theta', \theta'') \\ &\quad + \sum_{\kappa_n} \int_{S^2} d\theta''\, e^{-\kappa_n t'} M_{\kappa_n}(\theta', \theta'') \mathscr{J}(i\kappa_n, \theta \cdot \theta'', |x|, t), \end{aligned} \tag{4.13}$$

where M and M_κ are the kernels in the spectral function (3.35).

The properties of the kernel $\mathscr{U}$ of the generalized Gel'fand-Levitan equation (4.12) remain unexplored and little is known about the solvability of either (4.8) or (4.12). If they can be solved uniquely, the potential can be computed from their solutions by means of (3.9) or (3.13), respectively.

Summary

Let us now summarize the use of the generalized Gel'fand-Levitan equation for the solution of the inverse scattering problem. As stated in Theorem 4.1.3, the inverse spectral problem is generically solvable by the generalized Gel'fand-Levitan equation and it will generally lead to a nonlocal potential of the Kay-Moses type. Furthermore, this solution will have as its spectral function the function that served as its starting point.

Suppose, on the other hand, that the starting point is a given S matrix. One then solves the Wiener-Hopf factorization problem and obtains the Jost function. Generically this can be done. One then computes a spectral function from (3.35), obtains $r(x, y)$ by means of (4.6), solves the generalized Gel'fand-Levitan equation (4.8), and computes ϕ by (3.16). This will generally lead to a Schrödinger equation for ϕ with a nonlocal potential. The next step is to define ψ^+ by (3.2). This function will now, by construction, satisfy equation (1.61), and it satisfies the same Schrödinger equation as ϕ. However, the function ζ defined by (1.52) can in general not be expected to have the property (1.54) and may, therefore, not solve the Riemann-Hilbert problem that it should solve. As a result, the initially given S matrix can generally not be expected to be the S matrix that corresponds to the nonlocal potential obtained from the solution of the inverse problem *via* the generalized Gel'fand-Levitan equation.[1]

If, however, the potential V constructed by means of the generalized Gel'fand-Levitan equation is *local*, then we proceed to solve the Lippmann-Schwinger equation and compute the scattering solution $\psi^{+\prime}$ and its scattering amplitude A'. We solve the Wiener-Hopf factorization, thereby constructing the Jost function J', and hence a regular solution ϕ' as well as the Povzner-Levitan kernels h' and w'. Do the primed quantities equal the unprimed?

Lemma 3.2.2 tells us that $w = w'$ and thus $\phi = \phi'$ and $M = M'$. The part of the spectral function M that refers to $k^2 < 0$ allows us to infer the point spectrum and the kernels $M_\kappa(\theta, \theta')$; then (3.32) leads to a set of eigenfunctions $u_\kappa^a(x)$ and (1.29) to the corresponding characters. We now use the following result, to be proved below.

Lemma 4.1.4. *The Wiener-Hopf factorization problem posed by the equation $M(k)$ $= \widetilde{F}(-k)F(k) - \mathbb{1}$, in which F is required to have all the properties required of J^{-1} in Section 2.6, has at most one solution.*

This lemma ensures that $J' = J$ and hence $\psi^{+\prime} = \psi^+$, and thus finally, $A' = A$. The circle is therefore closed: The potential constructed is indeed the potential that underlies the given scattering amplitude.

---

[1] The question of whether a solution of the Schrödinger equation has S as its S matrix, or A as its scattering amplitude, in the asymptotic sense of equation (1.2) has to be distinguished from whether it has S as its S matrix in the sense of the Riemann-Hilbert problem. For a potential of the Kay-Moses type the scattering solution defined by (1.3) will generally not be analytic in $\mathbb{C}^+$, and the solution defined by (3.2) will in general not have the right large-$|k|$ asymptotics. However, the question whether the solution ψ^+ defined by (3.2) has the appropriate large-$|x|$ asymptotics with the correct scattering amplitude A that was used as input, is open. If it does, then the circle $A \longrightarrow V \longrightarrow A$ is closed for these nonlocal potentials, after all, even though the path *via* the Riemann-Hilbert problem does not work for them.

Theorem 4.1.5. *Suppose that the given scattering amplitude $A \in \mathcal{A}$ (see Definition 1.5.15), a corresponding Jost function exists, and the generalized Gel'fand-Levitan equation has a solution that produces a local potential $V \in \mathcal{W}$ (see Definition 1.4.1) by (3.13). Then this potential has A as its scattering amplitude.*

The only step missing in the above argument is the proof of Lemma 4.1.4, and it will now by supplied.

Proof of Lemma 4.1.4. Suppose $F^\dagger F = F'^\dagger F'$, so that we have $\tilde{J}'(-k)\tilde{F}(-k) = F'(k)J(k)$. The ranges of the residues of F at its poles are given as the spaces spanned by the respective characters, and these equal the nullspaces of J at the same points. Form the operators Π_r defined by (2.80), using real symmetric projections B_m. Then $F'J = \tilde{J}'^*\tilde{F}^*$, multiplied on the left by Π_r^{-1} and on the right by Π_r gives $\Gamma := F'^{\mathrm{red}} J^{\mathrm{red}} = \tilde{J}'^{\mathrm{red}*}\tilde{F}^{\mathrm{red}*}$, where $F^{\mathrm{red}} = \Pi_r^{-1}F$, $J^{\mathrm{red}} = J\Pi_r$, $F'^{\mathrm{red}} = \Pi_r^{-1}F'$, $J'^{\mathrm{red}} = J'\Pi_r$, and these are all holomorphic in $\mathbb{C}^+$. Therefore the function Γ is holomorphic both in $\mathbb{C}^+$ and in $\mathbb{C}^-$, and continuous on $\mathbb{R}$. Thus it is an entire analytic function that tends to $\mathbb{1}$ at infinity, and by Liouville's theorem $\Gamma = \mathbb{1}$. Therefore $F'J = \mathbb{1}$, and since both have inverses, $J = F'^{-1} = J'$. $\qquad\square$

4.2 An Example

The method of this chapter is the only one of this book that lends itself to the construction of simple solvable examples. The reason for that is, of course, that no miracle is required for the solution of the generalized Gel'fand-Levitan to make sense as a solution of an inverse problem.

Suppose that we assume for the spectral function in (3.35)

$$M(k) = 0, \qquad M_\kappa(\theta, \theta') = NX(\theta)X(\theta'),$$

where the function X, $S^2 \mapsto \mathbb{R}$, is arbitrary, but we assume that it is normalized so that $\|X\| = 1$; N is a real constant. In other words, there is a single nondegenerate bound state, and the spectral function for the continuous spectrum equals that for the free solution. Then by (4.6)

$$r(x, y) = NZ(x)Z(y),$$

where the function Z is defined by (3.24), and equation (4.8) becomes for $|y| < |x|$,

$$h(x, y) = NZ(x)Z(y) - NZ(y) \int_{\mathscr{B}(|x|)} dz\, Z(z)h(x, z).$$

Setting

$$\lambda(x) := \int_{\mathscr{B}(|x|)} dz\, Z(z)h(x, z),$$

one finds

$$\lambda(x) = \frac{Z(x)}{\Lambda(|x|)} \int_{\mathscr{B}(|x|)} dy\, Z^2(y),$$

where

$$A(|x|) = \frac{1}{N} + \int_{\mathscr{B}(|x|)} dy\, Z^2(y).$$

As a result the solution of (4.8) is

$$h(x, y) = \frac{Z(x)Z(y)}{A(|x|)},$$

and the potential, which is nonlocal in direction, is given by (4.9)

$$V(r, \hat{x}, \hat{x}') = -2\frac{\partial}{\partial r}\left[r^2\frac{Z(r\hat{x})Z(r\hat{x}')}{A(r)}\right]$$

in the sense that $(Vf)(r\hat{x}) = \int_{S^2} d\hat{x}'\, V(r, \hat{x}, \hat{x}')f(r\hat{x}')$. The regular solution of the Schrödinger equation is given by

$$\phi(k, \theta, x) = e^{ik\theta \cdot x} - \frac{Z(x)}{A(|x|)}\Gamma(k, \theta, |x|),$$

where

$$\Gamma(k, \theta, |x|) = \int_{\mathscr{B}(|x|)} dy\, Z(y)e^{ik\theta \cdot y}.$$

Because of (3.25), the bound-state eigenfunction is similarly given by

$$u(x) = \frac{Z(x)}{N A(|x|)}.$$

The asymptotic forms of these functions for large $|x| = r$ is readily found to be

$$Z(x) = \frac{2\pi}{\kappa r}X(-\hat{x})e^{\kappa r} + o\left(\frac{e^{\kappa r}}{r}\right),$$

$$A(r) = \frac{2\pi^2}{\kappa^3}e^{2\kappa r} + o(e^{2\kappa r}),$$

$$\Gamma(k, \theta, r) = -\frac{(2\pi)^2}{k\kappa}\left[X(-\theta)\frac{e^{ikr}}{k - i\kappa} + X(\theta)\frac{e^{-ikr}}{k + i\kappa}\right]e^{\kappa r} + o(e^{\kappa r}).$$

Using Corollary 1.5.2 one obtains the scattering amplitude from the asymptotic form of ϕ for large r,

$$A(k, \theta, \theta') = \frac{4\pi\kappa}{k(k - i\kappa)}\left[P(\theta, \theta') + P(-\theta, -\theta') + \frac{2i\kappa c}{k - i\kappa}P(-\theta, \theta')\right],$$

where $P(\theta, \theta') := X(\theta)X(\theta')$ is the kernel of the orthogonal projection onto the space spanned by X, and $c := (X, QX) = \int d\theta\, X(\theta)X(-\theta)$. The Jost function is easily recognized as the factor of the incoming-wave part of ϕ,

$$J = \mathbb{1} - \frac{2i\kappa}{k + i\kappa}P.$$

It satisfies equation (2.77). The scattering solution ψ^+, obtained from equation (3.2), fails to solve the Riemann-Hilbert problem because its asymptotic form for large $|k|$ in $\mathbb{C}^+$ does not obey Lemma 1.4.2.

It is clear that this example can easily be generalized to more than one bound state. The class of potentials thus generated is, however, not large enough to include any local ones.

4.3 Reduction to Central Potentials

Of all the methods discussed in this volume the only one that reduces to the known equations for solving the inverse scattering problem for central potentials is the generalized Gel'fand-Levitan method. Suppose that $V(x) = v(|x|)$. In that case the function $h(x, y)$ can depend on $\hat{x}$ and $\hat{y}$ only *via* $\hat{x} \cdot \hat{y}$ and we expand

$$|x| \, |y| h(x, y) = \sum_L Y_L(\hat{x}) \overline{Y}_L(\hat{y}) K_l(|x|, |y|)$$

in the notation for the spherical harmonics employed in Section 1.6. Equation (3.12) then becomes for $s, t \in \mathbb{R}_+$, $s < t$

$$\left[\frac{\partial^2}{\partial s^2} - \frac{\partial^2}{\partial t^2} + l(l+1) \left(\frac{1}{t^2} - \frac{1}{s^2} \right) + V(t) \right] K_l(t, s) = 0,$$

and $K_l(t, s) = 0$ for $s > t$. Equation (3.13) implies that for all l

$$-2 \frac{\partial}{\partial s} K_l(s, s) = v(s). \tag{4.14}$$

The functions $\phi_l(k, r)$ and $\phi_l^0(k, r)$ are defined by the expansions

$$\phi(k, \theta, x) = \sum_L \frac{(ik)^l}{|x|(2l+1)!!} \phi_l(k, |x|) \overline{Y_L(\theta)} Y_L(\hat{x}), \tag{4.15}$$

$$e^{ik\theta \cdot x} = \sum_L \frac{(ik)^l}{|x|(2l+1)!!} \phi_l^0(k, |x|) \overline{Y_L(\theta)} Y_L(\hat{x}).$$

Then the representation (3.16) leads to

$$\phi_l(k, r) = \phi_l^0(k, r) - \int_0^r ds \, K_l(r, s) \phi_l^0(k, s),$$

and the generalized Gel'fand-Levitan equation (4.8) is readily found to lead to the equation

$$K_l(u, s) = g_l(u, s) - \int_0^u dt \, K_l(u, t) g_l(t, s), \tag{4.16}$$

for $s < u$. Here g_l is defined by the expansion

$$|x| \, |y| r(x, y) = \sum_L Y_L(\hat{x}) \overline{Y_L(\hat{y})} g_l(|x|, |y|) / (2l+1)!!,$$

so that by (4.6)

$$g_l(t, s) = \int \left(\phi_l^0(k, t), \, d[\rho_l(k) - \rho_l^0(k)] \phi_l^0(k, s) \right),$$

where $d(\rho_l - \rho_l^0)$ is defined by the expansion

$$d(\rho - \rho^0)(k, \theta, \theta') = \sum_L Y_L(\theta) \overline{Y}_L(\theta') d(\rho_l - \rho_l^0)(k).$$

Equation (4.16) is the well-known Gel'fand-Levitan equation for the case of central potentials, and (4.14) is the equation that leads to the potential in that case. The miracle, for central potentials, is the fact that the left-hand side of (4.14) has to be independent of l.

The relation (3.2) between the regular solution and the scattering solution together with the expansions

$$\psi^+(k,\theta,x) = \frac{1}{k|x|}\sum_L i^l \psi_l^+(k,|x|)\overline{Y}_L(\theta)Y_L(\hat{x}),$$

$$J(k,\theta,\theta') = \sum_L f_l(k)Y_L(\theta)\overline{Y}_L(\theta'),$$

or in terms of Legendre polynomials

$$J(k,\theta,\theta') - \delta(\theta,\theta') = \frac{1}{4\pi}\sum_l (2l+1)[f_l(k)-1]P_l(\theta\cdot\theta'),$$

and (4.15) lead to

$$\phi_l(k,|x|) = (2l+1)!!k^{-l-1}f_l(k)\psi_l^+(k,|x|).$$

It may be concluded from this that $f_l(k)$ is the Jost function of the radial Schrödinger equation for the angular momentum l.

Thus all the three-dimensional equations in the Gel'fand-Levitan method reduce to the well-known ones for central potentials. The question whether h and $J - \mathbb{1}$ are functions or distibutions is related to the question of the convergence of the above expansions. That a similar reduction of the generalized Marchenko equation for central potentials does not appear to be possible should not come as a surprise because the radial Marchenko equations deal with Jost solutions of the Schrödinger equation, which have no three-dimensional analogue.

4.4 Notes

4.1 The content of this section is a generalization and refinement of parts of [Ne80, 81, and 82a]. Lemma 4.1.2 and Theorem 4.1.3 are new. The basic idea of the paper [KM61a] was excellent. However, as a result of a technical error (on the bottom of p.700) the authors obtained an incorrect spectral function. Furthermore, they did not realize that the nonlocal potential obtained does not necessarily lead back to the originally given S matrix, even though it gives the correct spectral function.

4.2 The example of this section is new.

4.3 This section follows [Ne82a].

5. Standing-Wave Solutions

5.1 The K-matrix

One of the standard solutions of the Schrödinger equation is defined by means of the Green's function

$$G_0^P(k, x) := \frac{1}{(2\pi)^3} \mathscr{P} \int_{\mathbb{R}^3} dk' \frac{e^{ik' \cdot x}}{k^2 - k'^2}$$

$$= -\frac{\cos k|x|}{4\pi|x|}, \tag{5.1}$$

where $\mathscr{P}$ stands for Cauchy's *principal value*, and the integral equation

$$\psi^P(k, \theta, x) = e^{ik\theta \cdot x} + \int_{\mathbb{R}^3} dy \, G_0^P(k, x - y) V(y) \psi^P(k, \theta, y). \tag{5.2}$$

This equation can be solved by Fredholm methods under the same conditions as the Lippmann-Schwinger equation. If $V \in \mathscr{V}$ (see Definition 1.1.1) Fredholm methods are applicable and the modified Fredholm determinant exists. Let us write

$$G_0^P = G_0^+ + i\mathscr{D},$$

where $\mathscr{D}(k)$ is the integral operator whose kernel is given by $\mathscr{D}(k, x - y)$,

$$\mathscr{D}(k, x) = \frac{k}{(4\pi)^2} \int_{S^2} d\theta \, e^{ik\theta \cdot x}. \tag{5.3}$$

One may factorize

$$\mathbb{1} - G_0^P V = \mathbb{1} - G_0^+ V - i\mathscr{D} V = (\mathbb{1} - G_0^+ V)\left(\mathbb{1} + \frac{ik}{4\pi}\mathscr{R}\right),$$

in which $\mathscr{R}$ is the integral operator with the kernel

$$\mathscr{R}(k, x, y) := -\frac{1}{4\pi} \int_{S^2} d\theta \, \psi^+(k, \theta, x) e^{-ik\theta \cdot y} V(y).$$

Now by the same argument used in the proof of Lemma 1.5.8 one finds that

$$\det_2[\mathbb{1} - G_0^P(k)V] = \Delta(k) \det\left[\mathbb{1} + \frac{ik}{4\pi}\mathscr{R}(k)\right]$$

$$\times \exp\left[-\text{tr}\left[(\mathbb{1} - G_0^+(k)V)\frac{ik}{4\pi}\mathscr{R}(k)\right]\right]$$

$$= \Delta(k) \det\left[\mathbb{1} - \frac{k}{4\pi i}A(k)\right] \exp\left(\frac{ik}{4\pi} \int_{\mathbb{R}^3} dx \, V(x)\right),$$

where $\varDelta(k)$ is the Fredholm determinant of the Lippmann-Schwinger equation, which does not vanish for real $k \neq 0$ if $V \in \mathscr{V}$. Therefore the Fredholm determinant of equation (5.2) fails to vanish for real $k \neq 0$ unless $(k/4\pi i)A(k)$ has the eigenvalue 1, which would mean that $S(k)$ has the eigenvalue -1.

Lemma 5.1.1. *If $V \in \mathscr{V}$ (see Definition 1.1.1), the solution ψ^P exists for all real $k \neq 0$ unless $S(k)$ has the eigenvalue -1.*

The function ψ^P is called the *principal-value solution* of the Schrödinger equation; it is a *standing-wave* solution. Both its real and imaginary parts are separately solutions, which, when multiplied by $e^{-i\omega t}$, makes them solutions of the time-dependent Schrödinger equation that are standing waves.

The above given factorization of $\mathbb{1} - G_0^P V$, when applied to ψ^P, gives

$$\psi^+(k,\theta,x) = \psi^P(k,\theta,x) + \frac{ik}{4\pi} \int_{S^2} d\theta'\, K(k,\theta',\theta)\psi^+(k,\theta',x), \tag{5.4}$$

where

$$K(k,\theta,\theta') := -\frac{1}{4\pi} \int_{\mathbb{R}^3} dx\, e^{-ik\theta \cdot x} V(x)\psi^P(k,\theta',x). \tag{5.5}$$

In a similar manner one finds that

$$\mathbb{1} - G_0^+ V = (\mathbb{1} - G_0^P V)\left(\mathbb{1} - \frac{ik}{4\pi}\mathscr{R}'\right),$$

where

$$\mathscr{R}'(k,x,y) := -\frac{1}{4\pi} \int_{S^2} d\theta\, \psi^P(k,\theta,x) e^{-ik\theta \cdot y} V(y),$$

and consequently

$$\psi^P(k,\theta,x) = \psi^+(k,\theta,x) - \frac{ik}{4\pi} \int_{S^2} d\theta'\, A(k,\theta',\theta)\psi^P(k,\theta',x). \tag{5.6}$$

Multiplying this equation by $V(x)e^{-ik\theta \cdot x}$ and integrating leads to *Heitler's integral equation*, which connects the function K to the scattering amplitude,

$$K(k,\theta,\theta') = A(k,\theta,\theta') - \frac{ik}{4\pi} \int_{S^2} d\theta''\, K(k,\theta,\theta'')A(k,\theta'',\theta'). \tag{5.7}$$

The *K matrix* or *reactance matrix* is defined to be the operator whose integral kernel is given by

$$\mathscr{K}(k,\theta,\theta') := \frac{k}{4\pi i} K(k,\theta,\theta'). \tag{5.8}$$

It follows from Heitler's integral equation that $\mathscr{K}$ is the Cayley transform of the S matrix

$$\mathscr{K} = (\mathbb{1} - S)(\mathbb{1} + S)^{-1}, \tag{5.9}$$

or

$$S = (\mathbb{1} - \mathscr{K})(\mathbb{1} + \mathscr{K})^{-1}. \tag{5.10}$$

The fact that G_0^P is real and symmetric implies that ψ^P has the symmetry

$$\psi^P(k, \theta, x) = \psi^P(-k, -\theta, x), \tag{5.11}$$

and that K has the symmetries

$$\overline{K}(k, \theta, \theta') = K(-k, \theta, \theta') = K(k, -\theta, -\theta'). \tag{5.12}$$

It also follows from equation (5.2) that K satisfies the reciprocity relation

$$K(k, \theta, \theta') = K(k, -\theta', -\theta). \tag{5.13}$$

Together, these equations imply that the integral operator K is *self-adjoint*, and thus $\mathcal{K}$ is *skew-adjoint*.

The function $\psi^-(k, \theta, x) := \psi^+(-k, -\theta, x)$, the incoming-wave solution of the Schrödinger equation, satisfies the Lippmann-Schwinger equation with G^+ replaced by G^-. The solution of the Schrödinger equation which, for real k, is the average of the two solutions ψ^+ and ψ^-,

$$\chi := \tfrac{1}{2}(\psi^+ + \psi^-) \tag{5.14}$$

is readily found by iteration of (5.4) and use of the symmetries (5.13) and (5.12) to be related to ψ^P by

$$\chi(k, \theta, x) = \psi^P(k, \theta, x) + \int_{S^2} d\theta' \, \chi(k, \theta', x) \mathcal{K}^2(k, \theta', \theta), \tag{5.15}$$

where $\mathcal{K}^2(k, \theta', \theta)$ is the kernel of the square of the operator $\mathcal{K}$. It also follows from (1.61) and (5.10) that ψ^+ and χ are connected by

$$\chi(k, \theta, x) = \psi^+(k, \theta, x) + \int_{S^2} d\theta' \, \chi(k, \theta', x) \mathcal{K}(k, \theta', \theta). \tag{5.16}$$

The solution χ of the Schrödinger equation can be expressed in terms of plane waves:

$$\chi(k, \theta, x) = e^{ik\theta \cdot x} + \int_{\mathbb{R}^3} dy \, G^P(k, x, y) V(y) e^{ik\theta \cdot y}, \tag{5.17}$$

where

$$G^P := \tfrac{1}{2}(G^+ + G^-)$$

is a complete Green's function. From its definition it is clear that χ has the symmetries

$$\chi(k, \theta, x) = \chi(-k, -\theta, x) = \overline{\chi}(k, \theta, x). \tag{5.18}$$

They imply that the solution $\chi(k, \theta, x) + \chi(k, -\theta, x)$ is real and thus also represents a standing wave.

5.2 The $\bar{\partial}$-method

Both the Marchenko procedure and the Gel'fand-Levitan method are based on the exploitation of analyticity properties of solutions of the Schrödinger equation. The $\bar{\partial}$-method enlarges the scope of the available tools to functions that are not analytic.

Let $f(x, y)$ be a continuous, differentiable complex-valued function of the two variables $x, y \in \mathbb{R}$. Such a function may always be expressed as a function of the two variables $z = x + iy$ and $\bar{z} = x - iy$, $f(x, y) = g(z, \bar{z})$. Now, f is *holomorphic* as a function of z if it satisfies the Cauchy-Riemann equations. These can be written compactly in the form

$$\frac{1}{2} \left(\frac{\partial}{\partial x} + i \frac{\partial}{\partial y} \right) f = \frac{\partial f}{\partial \bar{z}} := \bar{\partial} f = 0.$$

On the other hand, suppose that f is *not* holomorphic, that is, it is continuous and differentiable but does not satisfy the Cauchy-Riemann equations. Instead, suppose that in some region $\Omega \in \mathbb{C}$ with boundary $\partial \Omega$,

$$\frac{\partial f}{\partial \bar{z}} = u,$$

where u is a given function. This equation can be explicitly solved by the *Bochner-Martinelli* formula

$$f(z) = \frac{1}{2\pi i} \int_{\partial \Omega} d\zeta \, \frac{f(\zeta)}{\zeta - z} - \frac{1}{2\pi i} \int_{\Omega} d\bar{\zeta} \wedge d\zeta \, \frac{\bar{\partial} f(\zeta)}{\zeta - z}, \tag{5.19}$$

where $d\bar{\zeta} \wedge d\zeta = 2i d\zeta_1 d\zeta_2$ if $\zeta = \zeta_1 + i\zeta_2$. This is the generalization of Cauchy's integral formula, to which it reduces if f is holomorphic in Ω. It is proved by means of Stokes's theorem [see Kr82].

Since the operator $\bar{\partial}$ annihilates any holomorphic function we have

$$\begin{aligned} \bar{\partial} f(z) &= \frac{1}{2\pi i} \int_{\Omega} d\bar{\zeta} \wedge d\zeta \, \bar{\partial} f(\zeta) \bar{\partial}_z \frac{1}{z - \zeta} \\ &= \frac{1}{\pi} \int_{\Omega} d\zeta_1 d\zeta_2 \, \bar{\partial} f(\zeta) \bar{\partial}_z \frac{1}{z - \zeta}. \end{aligned}$$

This allows us to conclude that

$$\bar{\partial}_z \frac{1}{z - \zeta} = \pi \delta(z_1 - \zeta_1) \delta(z_2 - \zeta_2) := \pi \delta_{\mathbb{C}}(z - \zeta), \tag{5.20}$$

where δ denotes the Dirac distribution.

For example, suppose that

$$f(z) = \int_{-\infty}^{\infty} d\zeta \, \frac{g(\zeta)}{\zeta - z} \tag{5.21}$$

with $g \in L^1(\mathbb{R})$, so that f is holomorphic in $\mathbb{C}$ except for a cut along the real line (where it may be interpreted to have the value $f(x) = \lim_{\epsilon \downarrow 0} \frac{1}{2}[f(x+i\epsilon) + f(x-i\epsilon)]$, which is Cauchy's principal value of the integral in (5.21), but this is of no consequence). It then follows from (5.20) that

$$\bar{\partial}f(z) = -\pi\delta(y)g(x).$$

Here $z = x + iy$ and $2\pi ig(x)$ is the discontinuity of f at x across the real axis:

$$\bar{\partial}f(z) = \tfrac{1}{2}i\delta(y)\lim_{\epsilon\downarrow 0}[f(x + i\epsilon) - f(x - i\epsilon)]. \qquad (5.22)$$

These equations have generalizations to higher dimensions, but we shall not explicitly need them.

5.3 The Inverse Problem

Let us go back to the full Green's function for the Lippmann-Schwinger equation, with $k = k_1 + ik_2 \in \mathbb{C}$. Its expansion is given by (1.48), which may be written in the form

$$G(k, x, y)$$
$$= \frac{1}{32\pi^3} \int_{-\infty}^{\infty} dk'\, k' \int_{S^2} d\theta\, \frac{\psi^+(k', \theta, x)\overline{\psi^+}(k', \theta, y) + \psi^-(k', \theta, x)\overline{\psi^-}(k', \theta, y)}{k - k'}$$
$$+ \sum_{n,a} \frac{u^a_{\kappa_n}(x)u^a_{\kappa_n}(y)}{k^2 + \kappa_n^2},$$

if the functions $u^a_{\kappa_n}$ are chosen to be orthonormal. For $k \in \mathbb{C}^+$ it equals the analytic continuation of G^+, for $k \in \mathbb{C}^-$ it equals that of G^-. For $k \in \mathbb{R}$ it may be taken to equal the function G^P in (5.17), but this is irrelevant. Regarded as a function of $k \in \mathbb{C}$ it is not everywhere holomorphic.

We may calculate the derivative of G with respect to $\bar{k}$, which we shall call simply $\bar{\partial}G$, by means of the formula (5.20):

$$\bar{\partial}G(k, x, y)$$
$$= \delta(k_2)\frac{k}{32\pi^2} \int_{S^2} d\theta\, [\psi^+(k, \theta, x)\overline{\psi^+}(k, \theta, y) + \psi^-(k, \theta, x)\overline{\psi^-}(k, \theta, y)]$$
$$- \sum_{n,a} \frac{i\pi}{2\kappa_n} u^a_{\kappa_n}(x)u^a_{\kappa_n}(y)[\delta_{\mathbb{C}}(k - i\kappa_n) - \delta_{\mathbb{C}}(k + i\kappa_n)]. \qquad (5.23)$$

Consider the representation of the solution of the Lippmann-Schwinger equation (1.3)

$$\psi(k, \theta, x) := e^{ik\theta\cdot x} + \int_{\mathbb{R}^3} dy\, G(k, x, y)V(y)e^{ik\theta\cdot y}$$

for all $k \in \mathbb{C}$. Assuming that the potential $V \in \mathcal{V}$ this equation defines $\psi(k, \theta, x)$ so that it equals $\psi^+(k, \theta, x)$ in $\mathbb{C}^+$, $\psi^-(k, \theta, x)$ in $\mathbb{C}^-$. Its interpretation on $\mathbb{R}$ is arbitrary. Using the above result for $\bar{\partial}G$ one finds

$$\bar{\partial}\psi(k, \theta, x)$$
$$= -\delta(k_2)\frac{k}{8\pi} \int_{S^2} d\theta'\, \left[\psi^+(k, \theta', x)A(-k, \theta, \theta') + \psi^+(-k, \theta', x)A(k, -\theta, \theta')\right]$$
$$+ \sum_{n,a} \frac{i\pi}{2\kappa_n} u^a_{\kappa_n}(x)\left[\delta_{\mathbb{C}}(k + i\kappa_n)Y^a_{\kappa_n}(\theta) - \delta_{\mathbb{C}}(k - i\kappa_n)Y^a_{\kappa_n}(-\theta)\right].$$

One then uses (5.16) and (5.18), and the fact that (5.10), (5.12), (1.60), (1.61), and the reciprocity relation imply that

$$(1 - \mathcal{K})A^\dagger + Q(1 - \widetilde{\mathcal{K}})\tilde{A}Q = \frac{8\pi i}{k}\mathcal{K},$$

in which Q is the operator defined above Definition 1.5.15. One thus obtains

$$\bar{\partial}\psi(k,\theta,x) = -i\delta(k_2) \int_{S^2} d\theta'\, \chi(k,\theta',x)\mathcal{K}(k,\theta',\theta)$$
$$+ \sum_{n,a} \frac{i\pi}{2\kappa_n} u_{\kappa_n}^a(x) \left[\delta_{\mathbb{C}}(k + i\kappa_n) Y_{\kappa_n}^a(\theta) - \delta_{\mathbb{C}}(k - i\kappa_n) Y_{\kappa_n}^a(-\theta)\right]. \quad (5.24)$$

It is instructive to obtain the result (5.24) in another manner that makes no use of the explicit representation of the full Green's function. Differentiation of the Lippmann-Schwinger equation yields $(1 - G_0 V)\bar{\partial}\psi = \bar{\partial}G_0 V\psi$ or

$$\bar{\partial}\psi = (1 - G_0 V)^{-1}\bar{\partial}G_0 V\psi + \sigma, \quad (5.25)$$

where σ is a sum of complex Dirac distributions from the poles of ψ. The question is, what meaning should be given to the inverse $(1 - G_0 V)^{-1}$ and to ψ in this equation on the real axis of k?

To find the answer let us rewrite the integral in (5.24) using (5.5), (5.2), (5.15), (5.12), and (5.13):

$$\int_{S^2} d\theta'\, \chi(k,\theta',x)\mathcal{K}(k,\theta',\theta)$$

$$= \frac{ik}{(4\pi)^2} \int_{S^2} d\theta' \int_{\mathbb{R}^3} dy\, \chi(k,\theta',x)\overline{\psi^P}(k,\theta',y)V(y)e^{ik\theta\cdot y}$$
$$= \frac{ik}{(4\pi)^2} \int_{S^2} d\theta'\, \psi^P(k,\theta',x) \int_{\mathbb{R}^3} dy\, e^{-ik\theta'\cdot y}V(y)\chi(k,\theta,y),$$

which implies that

$$-i\delta(k_2) \int_{S^2} d\theta'\, \chi(k,\theta',x)\mathcal{K}(k,\theta',\theta) = (1 - G_0^P V)^{-1}\bar{\partial}G_0 V\chi.$$

This gives the precise meaning to be attached to the right-hand side of (5.25).

The idea now is to construct ψ from a knowledge of $\bar{\partial}\psi$ by means of the formula (5.19), in which Ω is taken to be $\mathbb{C}$. As previously in the case of ψ^+, it is first necessary to take out the exponential factor and to consider $\psi(k,\theta,x)e^{-ik\theta\cdot x}-1$. Since $\psi = \psi^+$ for $k_2 > 0$ and $\psi = \psi^-$ for $k_2 < 0$, it follows from Lemma 1.4.2 that if $V \in \mathcal{W}$ then in $\mathbb{C}^+$ or in $\mathbb{C}^-$,

$$\lim_{|k|\to\infty} [\psi(k,\theta,x)e^{-ik\theta\cdot x} - 1] = 0.$$

On the real axis it also follows from Lemma 1.4.2 that the analogue of (1.56) holds. Thus (5.19) may be applied, with the first term absent if $\Omega = \mathbb{C}$:

$$\psi(k,\theta,x)e^{-ik\theta\cdot x} - 1 = \frac{1}{\pi} \int_{\mathbb{R}^2} dk_1'dk_2' \frac{e^{-ik'\theta\cdot x}}{k - k'}\bar{\partial}\psi(k',\theta,x).$$

The left-hand side becomes $\psi^+ e^{-ik\theta \cdot x} - 1$ for $k \in \mathbb{C}^+$, and $\psi^- e^{-ik\theta \cdot x} - 1$ for $k \in \mathbb{C}^-$. The interpretation of this equation for $k \in \mathbb{R}$ is arbitrary. If it is interpreted as the average of its limits from above and below, then the integral on the right becomes Cauchy's principal value and ψ on the left equals χ, the function defined in (5.14). One then obtains by (5.24)

$$\chi(k, \theta, x)e^{-ik\theta \cdot x} - 1 = \frac{i}{\pi} \mathscr{P} \int_{-\infty}^{\infty} dk' \frac{e^{-ik'\theta \cdot x}}{k' - k} \int_{S^2} d\theta' \, \chi(k', \theta', x) \mathscr{K}(k', \theta', \theta)$$

$$+ \sum_{n,a} \frac{i}{2\kappa_n} u_{\kappa_n}^a(x) \left[\frac{e^{-\kappa_n \theta \cdot x}}{k + i\kappa_n} Y_{\kappa_n}^a(\theta) - \frac{e^{\kappa_n \theta \cdot x}}{k - i\kappa_n} Y_{\kappa_n}^a(-\theta) \right]. \qquad (5.26)$$

This equation, which is due to Somersalo [So87], is the analogue of equation (2.2) with the reactance matrix as the input, rather than the scattering amplitude.

The potential is obtained by inserting (5.26) in the Schrödinger equation and using the fact that χ solves that equation with the potential V. One finds

$$V(x)\chi(k, \theta, x)e^{-ik\theta \cdot x}$$

$$= -\frac{i}{\pi} \int_{-\infty}^{\infty} dk' \int_{S^2} d\theta' \left[2i\theta \cdot \nabla + \frac{V(x)}{k - k'} \right] e^{-ik'\theta \cdot x} \chi(k', \theta', x) \mathscr{K}(k', \theta', \theta)$$

$$+ \sum_{n,a} \frac{i}{2\kappa_n} \left[\left[\frac{V(x)}{k + i\kappa_n} + 2i\theta \cdot \nabla \right] e^{-\kappa_n \theta \cdot x} Y_{\kappa_n}^a(\theta) \right.$$

$$\left. - \left[\frac{V(x)}{k - i\kappa_n} + 2i\theta \cdot \nabla \right] e^{\kappa_n \theta \cdot x} Y_{\kappa_n}^a(-\theta) \right] u_{\kappa_n}^a(x),$$

the limit of which for large k gives the following expression for $V(x)$:

$$V(x) = 2\theta \cdot \nabla \left[\frac{1}{\pi} \int_{-\infty}^{\infty} dk' \int_{S^2} d\theta' \, e^{-ik'\theta \cdot x} \chi(k', \theta', x) \mathscr{K}(k, \theta', \theta) \right.$$

$$\left. - \sum_{n,a} \frac{1}{2\kappa_n} u_{\kappa_n}^a(x) \left[Y_{\kappa_n}^a(\theta) e^{-\kappa_n \theta \cdot x} - Y_{\kappa_n}^a(-\theta) e^{\kappa_n \theta \cdot x} \right] \right]. \qquad (5.27)$$

The miracle, of course, is again that the right-hand side has to be independent of θ. This representation of the potential is identical to that given previously in (2.105), except that it is expressed in terms of the reactance matrix instead of the scattering amplitude.

5.4 Notes

5.1 The material of this section is well known; see, for example [Ne82c].

5.2 References for the $\bar{\partial}$-method are the books by Hörmander [Ho66] and Krantz [Kr82].

5.3 This section is based on [So87], though we proceed somewhat differently from Somersalo. Equation (5.26) is due to Somersalo, but his version differs because he went astray in the proper interpretation of (5.25), and he assumed that there are no bound states. As a result, Somersalo's formula for the potential also differs from (5.27).

Part III

Use of the Faddeev Solution

6. Faddeev's Solution

6.1 The Green's Function

The solutions of the Schrödinger equation on which the methods discussed up to this point were based are all ultimately defined by the Lippmann-Schwinger equation. The Green's function contained in that integral equation is the kernel of the resolvent of the Laplacian on $L^2(\mathbb{R}^3)$. We now turn to procedures that are based on the use of a family of Green's functions introduced by Faddeev, which may be regarded as resolvent kernels of the Laplacian on weighted L^2-spaces (on which the Laplacian is not essentially self-adjoint). The Green's function (1.4) is one member of this family. There is a variety of ways of approaching these Green's functions, but we shall do so by an avenue originally used by Faddeev.

Consider the function $e^{iK\cdot x}$ with $K = k + iq \in \mathbb{C}^3$, $k, q \in \mathbb{R}^3$; it is a solution of the Helmholtz equation $(\Delta + K^2)e^{iK\cdot x} = 0$, and for fixed q the functions $e^{iK\cdot x}$, with k ranging over $\mathbb{R}^3$, span the weighted space $L^2(\mathbb{R}^3, w)$, $w = e^{2q\cdot x}$. The spectrum of the Laplacian on this space consists of the interior of the parabola $\lambda = \lambda_1 + i\lambda_2$, $\lambda_1 = q^2 - \lambda_2^2/4q^2$. In the sense of L^2-Fourier integrals we have on $L^2(\mathbb{R}^3, w)$

$$\frac{1}{(2\pi)^3} \int_{\mathbb{R}^3} dp \, e^{i(p+iq)\cdot(x-y)} = \delta(x - y).$$

Consequently the kernel of $(\Delta + \lambda)^{-1}$ is given by

$$F_q(\lambda, x - y) = \frac{1}{(2\pi)^3} \int_{\mathbb{R}^3} dp \, \frac{e^{i(p+iq)\cdot(x-y)}}{\lambda - (p + iq)^2},$$

(where we have used the notation $k^2 := k \cdot k$ for $k \in \mathbb{C}^3$) so that

$$(\Delta + \lambda)F_q(\lambda, x - y) = \delta(x - y).$$

Let us then take a vector $K = k + iq \in \mathbb{C}^3\backslash\mathbb{R}^3$ so that $K^2 = \lambda$. Defining $q = t\gamma$, with $\gamma \in \mathbb{R}^3$, $|\gamma|^2 = 1$, and $t > 0$, we decompose it as $K = k_\perp + \gamma s$, where $k_\perp \cdot \gamma = 0$, $s = k_0 + it \in \mathbb{C}^+$, $k_0 = k \cdot \gamma$. For given $\lambda = \lambda_1 + i\lambda_2 \in \mathbb{C}$ and $q \in \mathbb{R}^3\backslash 0$ we have $\lambda_2 = 2k_0 t$, $\lambda_1 = \mu^2 + k_0^2 - t^2$, where $\mu := |k_\perp|$. Thus, except for the direction of $k_\perp$ in the plane orthogonal to γ, K is uniquely determined and we may write $F_q(\lambda, x) := F_\gamma(\mu, s, x) := \tilde{F}(K, x)$. A shift in the variables of integration according to $p' = p - \gamma k_0$ gives us the representation

$$\hat{F}(k_\perp + \gamma s, x) = F_\gamma(\mu, s, x) = \frac{1}{(2\pi)^3} \int_{\mathbb{R}^3} dp' \, \frac{e^{ip' \cdot x + isx_0}}{\mu^2 - p'^2 - 2sp'_0}, \tag{6.1}$$

where $x_0 := x \cdot \gamma$, $p_0 := p \cdot \gamma$.

It is clear that F_γ is an analytic function of s that is holomorphic in $\mathbb{C}^+$. Of course, it was a matter of definition that we chose $t > 0$. Had we chosen $t < 0$, the direction of γ would have been reversed and, for the same given vector $K = k + iq$, s would have been replaced by $-s \in \mathbb{C}^-$. Thus we have for $s \notin \mathbb{R}$

$$F_\gamma(\mu, s, x) = F_{-\gamma}(\mu, -s, x). \tag{6.2}$$

The function F_γ, regarded as a function of $s \in \mathbb{C}$, has a branch cut along the entire real axis.[1] Let us denote its limits from above and below by F_γ^+ and F_γ^-, respectively. We then have from (6.2) for $s \in \mathbb{R}$,

$$F_\gamma^+(\mu, s, x) := F_\gamma(\mu, s + i0, x) = F_{-\gamma}(\mu, -s - i0, x) := F_{-\gamma}^-(\mu, -s, x),$$

and therefore its discontinuity across the real axis is given by

$$F_\gamma^+(\mu, s, x) - F_\gamma^-(\mu, s, x) = F_\gamma(\mu, s + i0, x) - F_{-\gamma}(\mu, -s + i0, x). \tag{6.3}$$

Note that the boundary values of $\hat{F}$ on $\mathbb{R}^3$ depend on the direction in which the boundary is approached. We shall therefore write for $k \in \mathbb{R}^3$

$$\hat{F}_\gamma^\pm(k, x) := \lim_{\epsilon \downarrow 0} \hat{F}(k \pm i\epsilon\gamma, x).$$

In Section 6.2.1 we shall prove the following two representations of the Faddeev Green's function. The first holds for $s \in \mathbb{C}$: For $\Im s \geq 0$

$$F_\gamma(\mu, s, x) = -\frac{e^{isr}}{4\pi r} + \frac{\mu}{4\pi} \int_{\hat{x} \cdot \gamma}^1 du \, \frac{e^{isru}}{\sqrt{1 - u^2}} J_1\left(\mu r \sqrt{1 - u^2}\right), \tag{6.4}$$

and for $\Im s \leq 0$

$$F_\gamma(\mu, s, x) = -\frac{e^{-isr}}{4\pi r} + \frac{\mu}{4\pi} \int_{-\hat{x} \cdot \gamma}^1 du \, \frac{e^{-isru}}{\sqrt{1 - u^2}} J_1\left(\mu r \sqrt{1 - u^2}\right), \tag{6.5}$$

where $\hat{x} = x/|x|$ and $r = |x|$. The second holds for real s:

$$\hat{F}_\gamma^\pm(k, x) = G_0^+(|k|, x) + B_{\pm\gamma}^0(k, x), \tag{6.6}$$

where

$$\begin{aligned} B_\gamma^0(k, x) &= \frac{i}{(2\pi)^2} \int_{\mathbb{R}^3} dp \, e^{ip \cdot x} \Theta[\gamma \cdot (p - k)] \delta(k^2 - p^2) \\ &= \frac{i|k|}{8\pi^2} \int_{\theta \cdot \gamma > \hat{k} \cdot \gamma} d\theta \, e^{i|k|\theta \cdot x}, \end{aligned}$$

---

[1] Note the difference between equations (6.2) and (3.3) and the different conclusions they entail, even though they look similar. Equation (3.3) holds on the real axis for the boundary value of a function that is analytic in $\mathbb{C}^+$. Equation (6.2), on the other hand, when taken to the real axis, relates the boundary values from below and above. Thus the first allows us to conclude, by Schwarz's reflection principle, that ϕ is an entire analytic function, but the second does not lead to an analogous conclusion.

and Θ is the Heaviside function

$$\Theta(u) = \begin{cases} 1, & \text{if } u > 0, \\ 0, & \text{if } u < 0. \end{cases}$$

The inequality

$$|J_1(u)| \le C \frac{u}{(1+u)^{3/2}}$$

for $u \ge 0$ easily leads to the following estimates:

$$\left| [F_\gamma(\mu, s, x) - G_0(s, x)] e^{-isx_0} \right| \le \begin{cases} C\mu^{1/2} r^{-1/2}, \\ C\mu, \\ C\mu^2/t, \end{cases} \tag{6.7}$$

if $s = k_0 + it$, $t > 0$. Here C is independent of x.

The split-ups (6.4) and (6.5) also directly show that the Faddeev Green's function has the following symmetries:

$$F_\gamma(\mu, -\bar{s}, x) = \overline{F_\gamma(\mu, s, x)}, \tag{6.8}$$

and

$$F_\gamma(\mu, s, x) = F_{-\gamma}(\mu, s, -x). \tag{6.9}$$

Remark. Equation (6.4) shows that the function $F_\gamma(\mu, s, x)$ has an analytic continuation as a function of s from the upper into the lower half plane. (This continuation is, of course, not equal to the function given by (6.5).) If $s = s_1 - it$, $t > 0$, then the continuation grows like e^{tr} as a function of x. Similarly, *mutatis mutandis*, for the analytic continuation of the function given by (6.5) into the upper half plane.

It is clear from the above split-ups that as $\mu \to 0$ the Faddeev Green's function goes over into the Green's function of the Lippmann-Schwinger equation. For $s \in \mathbb{R}$

$$F_\gamma^\pm(0, s, x) = G_0^\pm(s, x) \tag{6.10}$$

and their analytic continuations into $\mathbb{C}^+$ and $\mathbb{C}^-$, respectively.

Sometimes it is more convenient to work with the function

$$\mathscr{F}_\gamma(k_\perp, s, x) = \widehat{\mathscr{F}}(k, x) := \widehat{F}(k, x) e^{-ik \cdot x}$$

$$= -\frac{1}{(2\pi)^3} \int_{\mathbb{R}^3} dp \frac{e^{ip \cdot x}}{p^2 + 2k \cdot p}. \tag{6.11}$$

The last form is obtained from (6.1) by a shift of the variables of integration. It is readily found, again by a shift of the integration variables, that if $p \in \mathbb{R}^3$ and such that $p^2 + 2p \cdot k$, then

$$\widehat{\mathscr{F}}(k + p, x) = e^{-ip \cdot x} \widehat{\mathscr{F}}(k, x). \tag{6.12}$$

For the original Green's function this means that if $k - k' \in \mathbb{R}^3$ and $k^2 = k'^2$ then

$$\widehat{F}(k', x) = \widehat{F}(k, x). \tag{6.13}$$

Let us now look at the asymptotic behavior of the Faddeev Green's function for large $r = |x|$ and real s, which is obtained from (6.6):

$$F_\gamma(\mu, s, x) = -\frac{1}{4\pi r}\left[e^{i|k|r}\Theta(sr - |k|x_0) + e^{-i|k|r}\Theta(-sr - |k|x_0)\right] + o(r^{-1}). \quad (6.14)$$

For the interpretation of this result it is helpful to define two sets of cones as follows.

Definition 6.1.1. For $k, \gamma \in \mathbb{R}^3$, $|\gamma| = 1$

$$\mathscr{C}^\pm_{\gamma,k} := \{x \in \mathbb{R}^3 \mid \pm x_0 > 0, \ |x_\perp|/|x_0| < \mu/|s|\},$$

where $x_0 = \gamma \cdot x$, $x_\perp = x - \gamma\gamma \cdot x$, $s = \gamma \cdot k$, $\mu = \sqrt{k^2 - s^2}$.

The asymptotic behavior expressed in (6.14) can be described by

$$F_\gamma^+(\mu, s, x) = \begin{cases} o(r^{-1}), & \text{if } x \in \mathscr{C}^+_{\gamma,k}, \\ -\dfrac{\cos|k|r}{2\pi r} + o(r^{-1}), & \text{if } x \in \mathscr{C}^-_{\gamma,k}, \\ -\dfrac{e^{i|k|r}}{4\pi r} + o(r^{-1}), & \text{if } x \notin \mathscr{C}^+_{\gamma,k} \cup \mathscr{C}^-_{\gamma,k}, \ s > 0, \\ -\dfrac{e^{-i|k|r}}{4\pi r} + o(r^{-1}), & \text{if } x \notin \mathscr{C}^+_{\gamma,k} \cup \mathscr{C}^-_{\gamma,k}, \ s < 0. \end{cases} \quad (6.15)$$

6.2 The Integral Equation

The Green's function defined in the last section can now be used to set up an integral equation for a solution of the Schrödinger equation. This solution will be denoted by $\varphi_\gamma(k_\perp, s, x) = \widehat{\varphi}(k, x)$. For $k \in \mathbb{C}^3 \setminus \mathbb{R}^3$

$$\widehat{\varphi}(k, x) = e^{ik\cdot x} + \int_{\mathbb{R}^3} dy\, \widehat{F}(k, x - y)V(y)\widehat{\varphi}(k, y). \quad (6.16)$$

It will be convenient to factor out the exponential and to define an analogue of the function ζ of Chapter 1:

$$\widehat{\xi}(k, x) = \xi_\gamma(k_\perp, s, x) := \widehat{\varphi}_\gamma(k, x)e^{-ik\cdot x}, \quad (6.17)$$

which satisfies the differential equation

$$(\Delta + 2ik \cdot \nabla)\xi = V\xi \quad (6.18)$$

and the integral equation

$$\widehat{\xi}(k, x) = 1 + \int_{\mathbb{R}^3} dy\, \widehat{\mathscr{F}}(k, x - y)V(y)\widehat{\xi}(k, y), \quad (6.19)$$

where $\widehat{\mathscr{F}}$ is defined by (6.11).

As in the case of the Lippmann-Schwinger equation it requires a transformation to cast this into the form of a manageable integral equation. Multiply (6.19) by $|V|^{1/2}$, define $\xi'_\gamma(k_\perp, s, x) := |V|^{1/2}(x)\xi_\gamma(k_\perp, s, x)$, and

$$Y_\gamma(|k_\perp|, s, x, y) := |V|^{1/2}(x)\mathscr{F}_\gamma(k_\perp, s, x - y)V^{1/2}(y),$$

where $V^{1/2}$ was defined below equation (1.7). The resulting integral equation is

$$\xi'_\gamma(k_\perp, s, x) = |V|^{1/2}(x) + \int_{\mathbb{R}^3} dy\, Y_\gamma(k_\perp, s, x, y)\xi'_\gamma(k_\perp, s, y). \quad (6.20)$$

Lemma 6.2.1. *If $V \in \mathcal{V}$ then for all $\gamma \in \mathbb{R}^3$, $|\gamma| = 1$, and all $|k_\perp| \geq 0$, $s \in \mathbb{C}$, the operator defined by the kernel Y_γ is in the Hilbert-Schmidt class.*

The simple proof of this lemma is based on the estimates (6.7) and will not be given in detail.

As a result, the solution ξ'_γ exists as an L^2-function of $x \in \mathbb{R}^3$ for all $\gamma \in \mathbb{R}^3$, $|\gamma| = 1$ and $k \in \mathbb{C}^3$, except at those *exceptional points* where the modified Fredholm determinant

$$
\begin{aligned}
D_\gamma(\mu, s) &= \hat{D}(k) := \det_2[\mathbb{1} - Y_\gamma(\mu, s)] \\
&= \det_2[\mathbb{1} - F_\gamma(\mu, s)V] = \det_2[\mathbb{1} - \widehat{\mathscr{F}}(k)V]
\end{aligned}
\tag{6.21}
$$

vanishes. It follows from (6.8), (6.9), and the reality of the potential that this Fredholm determinant has the symmetries

$$
D_\gamma(\mu, -\bar{s}) = \overline{D_\gamma(\mu, s)} \ ,
\tag{6.22}
$$

$$
D_\gamma(\mu, s) = D_{-\gamma}(\mu, s).
\tag{6.23}
$$

Equation (6.10) implies that as $\mu \to 0$, D_γ approaches the Fredholm determinant of the Lippmann-Schwinger equation,

$$
D_\gamma(0, s) = \Delta(s).
\tag{6.24}
$$

If ξ'_γ is inserted in the right-hand side of (6.16), it is seen that φ_γ and ξ_γ exist pointwise and are continuous functions of s except at the exceptional points. In fact, one readily proves that the operator Y_γ is an analytic function of s that is holomorphic everywhere in $\mathbb{C}$, except on the real line, and the same holds for the function $D_\gamma(\mu, s)$. The functions $\varphi_\gamma(k_\perp, s, x)$ and $\xi_\gamma(k_\perp, s, x)$, on the other hand, are meromorphic in $\mathbb{C}^+$, with poles at the exceptional points; the product $D_\gamma(\mu, s)\varphi_\gamma(k_\perp, s, x)$ is an analytic function of s that is holomorphic in all of $\mathbb{C}^+$ and has a continuous boundary value on the real axis. It follows from the remark after equation (6.7) that if the potential decreases exponentially, i.e., if it satisfies the condition (1.80) for some $\epsilon > 0$, then $\varphi_\gamma(k_\perp, s, x)$ has an analytic continuation into $\mathbb{C}^-$ up to $\Im s > -\epsilon$. This implies particularly that if $\exists \epsilon > 0$ such that (1.80) holds then $\varphi_\gamma(k_\perp, s, x)$ is analytic *on the real axis* as a function of s. For large $|s|$ in $\mathbb{C}^+$, Y_γ vanishes. The following will be proved in Section 6.2.1.

Lemma 6.2.2. *If $V \in L^2(\mathbb{R}^3)$ then the kernel Y_γ is such that as either $s_1 := \Re s \to \pm\infty$ for $s_2 := \Im s > 0$, or as $s_2 \to \infty$,*

$$
\lim \ \|Y_\gamma(\mu, s, \cdot, \cdot)\|_2 = 0,
$$

where $\| \cdot \|_2$ is the Hilbert-Schmidt norm. The first limit is uniform in s_2 and the second in s_1.

The following is an immediate consequence, the last part as a result of equation (6.20).

Corollary 6.2.3. *Under the same conditions and in the same sense as in Lemma 6.2.2,*

$$
\lim \ D_\gamma(\mu, s) = 1,
$$

as well as

$$\lim \ \|\xi_\gamma'(k_\perp, s, \cdot) - |V|^{1/2}(\cdot)\| = 0.$$

In order to draw conclusions from this for the function ξ_γ it is convenient to rewrite equation (6.16) as follows:

$$\begin{aligned}
[\xi_\gamma(k_\perp, s, x) &- 1] \\
&= \int_{\mathbb{R}^3} dy \ [F_\gamma(\mu, s, x - y) - G_0(s, x - y)] e^{ik \cdot (y-x)} V^{1/2}(y) \\
&\quad \times [\xi_\gamma'(k_\perp, s, y) - |V|^{1/2}(y)] \\
&\quad - \frac{1}{4\pi} \int_{\mathbb{R}^3} dy \ \frac{e^{is[|y-x|+(y_0-x_0)]}}{|x - y|} V^{1/2}(y) [\xi_\gamma'(k_\perp, s, y) - |V|^{1/2}(y)] e^{k_\perp \cdot (y_\perp - x_\perp)} \\
&\quad + \int_{\mathbb{R}^3} dy \ [F_\gamma(\mu, s, x - y) - G_0(s, x - y)] e^{ik \cdot (y-x)} V(y) \\
&\quad - \frac{1}{4\pi} \int_{\mathbb{R}^3} dy \ \frac{e^{is[|y-x|+(y_0-x_0)]}}{|x - y|} V(y) e^{ik_\perp \cdot (y_\perp - x_\perp)}.
\end{aligned}$$

The use of the estimate (6.7) and of Corollary 6.2.3 allows us to conclude

Lemma 6.2.4. *If $V \in \mathcal{V}$ then for $s \in \mathbb{C}^+$ with $0 < \arg s < \pi$,*

$$\lim_{|s| \to \infty} \ \|\xi_\gamma(k_\perp, s, \cdot) - 1\| = 0.$$

The situation on the real axis is somewhat more complicated. There the split-up (6.6) is very useful. One then has the following result, which will be proved in Section 6.2.1.

Lemma 6.2.5. *Assume that $V \in L^1(\mathbb{R}^3)$ and that $\exists V_1, \ \mathbb{R}_+ \mapsto \mathbb{R}_+, \ \int_0^\infty dt \, t V_1(t) < \infty$, $\int_0^\infty dt \, t V_1^2(t) < \infty$, such that $|v(p)| < V_1(|p|)$, where v is the Fourier transform of $|V|$. Then the operator family $T(k), \ k \in \mathbb{R}^3$, defined by the integral kernels*

$$T(k, x, y) := |V|^{1/2}(x) V^{1/2}(y) B_\gamma^0(k, x - y),$$

has the property that $\exists C < \infty$ such that for all $k \in \mathbb{R}^3$, $\|T(k)\|_2 < C$. Furthermore,

$$\lim_{|k| \to \infty} \ \|T(k) T^\dagger(k)\|_2 = 0.$$

Here $\| \cdot \|_2$ denotes the Hilbert-Schmidt norm.

This lemma, together with Lemma 6.2.9 of Section 6.2.1, implies the following, by (6.6):

Lemma 6.2.6. *If V satisfies the hypotheses of Lemma 6.2.5 and it is in the Rollnik class, then $\exists C$ such that for all $\mu, \epsilon \in \mathbb{R}_+, \ s \in \mathbb{R}, \ \|\Upsilon_\gamma(\mu, s)\|_2 < C$ and*

$$\lim_{\mu^2 + s^2 \to \infty} \ \|\Upsilon_\gamma(\mu, s) \Upsilon_\gamma^\dagger(\mu, s)\|_2 = 0.$$

For the Fredholm determinant the corresponding result is the following, which is proved in Section 6.2.1

Lemma 6.2.7. *If $V \in \mathscr{V}$ then for $s, \mu \in \mathbb{R}_+$*

$$\lim_{s^2+\mu^2 \to \infty} \arg D_\gamma(\mu, s) = 0.$$

The boundary values of φ_γ as s approaches the real axis from above and from below will be denoted by $\varphi_\gamma^+(k_\perp, s, x) = \widehat{\varphi}_\gamma^+(k, x)$ and $\varphi_\gamma^-(k_\perp, s, x) = \widehat{\varphi}_\gamma^-(k, x)$, respectively; similarly for ξ_γ. It follows from (6.2) that

$$\varphi_\gamma^-(k_\perp, s, x) = \varphi_{-\gamma}^+(k_\perp, -s, x), \tag{6.25}$$

which may also be written

$$\widehat{\varphi}_\gamma^-(k, x) = \widehat{\varphi}_{-\gamma}^+(k, x). \tag{6.26}$$

Similarly, from (6.8),

$$\overline{\varphi_\gamma^+(k_\perp, s, x)} = \varphi_\gamma^+(-k_\perp, -s, x), \tag{6.27}$$

or

$$\overline{\widehat{\varphi}_\gamma^+(k, x)} = \widehat{\varphi}_\gamma^+(-k, x). \tag{6.28}$$

The function ξ_γ has the same symmetries.

6.2.1 Proofs

This section will contain derivations of (6.4), (6.5), and (6.6), as well as proofs of Lemmas 6.2.2, 6.2.7, and 6.2.5.

To get (6.4) and (6.5), we start with the representation (6.1),

$$F_\gamma(\mu, s, x) = \frac{1}{(2\pi)^3} \int_{\mathbb{R}^3} dp \, \frac{e^{ip \cdot x + isx_0}}{\mu^2 - p^2 - 2sp_0}.$$

For $\pm s > 0$ we insert the representation

$$\frac{1}{\mu^2 - p^2 - 2sp_0} = \frac{\pm i}{2p_0} \int_0^\infty dw \, \exp\left[\pm i\left(s - \frac{\mu^2 - p^2}{2p_0}\right)w\right].$$

The integral over the plane perpendicular to γ becomes

$$\int_{\mathbb{R}^2} dp_\perp \exp\left[ip_\perp \cdot x_\perp + i\frac{w}{2p_0}p_\perp^2\right] = \frac{2\pi i p_0}{w} \exp\left[-i\frac{p_0}{2w}x_\perp^2\right],$$

and two changes of variables, first setting $p_0 = wl/r$ and then $w = ru - x_0$, where $r = |x|$, leads to

$$F_\gamma(\mu, s, x) = \mp\frac{1}{8\pi^2} \int_{x_0/r}^{\pm\infty} du \int_{-\infty}^\infty dl \, \exp\left[ir[su + \tfrac{1}{2}l(u^2 - 1) - \tfrac{1}{2}\mu^2/l]\right].$$

Now note that

$$\int_{-\infty}^\infty dz \, e^{i(az+b/z)} = \begin{cases} 0, & \text{if } ab < 0, \\ -2\pi\sqrt{b/a}J_1(2\sqrt{ab}), & \text{if } ab > 0, \end{cases}$$

where J_1 is a Bessel function, by the use of a Mehler-Sonine formula and a change of integration variables. Therefore the u-integral terminates at ± 1. The neighborhoods of ± 1 have to be handled separately:

$$-\frac{1}{8\pi^2} \int_{1-\epsilon}^{1+\epsilon} du \int_{-\infty}^{\infty} dl\, e^{-ir[\frac{1}{2}l(1-u^2)+\frac{1}{2}\mu^2/l]}$$

$$= -\frac{\epsilon}{8\pi^2} \int_{-1}^{1} dv \int_{-\infty}^{\infty} dl\, e^{-ir(l\epsilon v+\frac{1}{2}\mu^2/l)}$$

$$= -\frac{1}{4\pi^2 r} \int_{-\infty}^{\infty} dl\, \frac{\sin(rl\epsilon)}{l} e^{-i\frac{1}{2}r\mu^2/l}$$

$$= -\frac{1}{4\pi^2 r} \int_{-\infty}^{\infty} dl\, \frac{\sin l}{l} e^{-i\frac{1}{2}r^2\epsilon\mu^2/l}$$

$$\xrightarrow{\epsilon \to 0} -\frac{1}{4\pi^2 r} \int_{-\infty}^{\infty} dl\, \frac{\sin l}{l} = -\frac{1}{4\pi r},$$

and the same result is obtained for the integral from $-1-\epsilon$ to $-1+\epsilon$. Therefore we find (6.4) and (6.5).

Equation (6.6) is derived as follows. As $\epsilon \downarrow 0$,

$$
\begin{aligned}
\widehat{F}(k+i\epsilon\gamma, x) &= \frac{1}{(2\pi)^3} \int_{\mathbb{R}^3} dp\, \frac{e^{ip\cdot x}}{(k+i\epsilon\gamma)^2 - (p+i\epsilon\gamma)^2} \\
&= \frac{1}{(2\pi)^3} \int_{\mathbb{R}^3} dp\, \frac{e^{ip\cdot x}}{k^2 - p^2 + 2i\epsilon\gamma \cdot (k-p)} \\
&= -\frac{i}{(2\pi)^3} \int_{\mathbb{R}^3} dp\, e^{ip\cdot x} \left[\Theta[\gamma \cdot (k-p)] \int_0^{\infty} dw\, e^{iw(k^2-p^2)} \right. \\
&\qquad\qquad\qquad\qquad \left. - \Theta[\gamma \cdot (p-k)] \int_{-\infty}^0 dw\, e^{iw(k^2-p^2)} \right] \\
&= -\frac{i}{(2\pi)^3} \int_{\mathbb{R}^3} dp\, e^{ip\cdot x} \int_0^{\infty} dw\, e^{iw(k^2-p^2)} \\
&\qquad + \frac{i}{(2\pi)^2} \int_{\mathbb{R}^3} dp\, e^{ip\cdot x} \Theta[\gamma \cdot (p-k)] \delta(k^2 - p^2).
\end{aligned}
$$

This gives (6.6).

Proof of Lemma 6.2.2.

Lemma 6.2.8. *The integral*

$$\mathscr{I} := \int_{\mathbb{R}^3} dp\, \frac{1}{(p^2 - \mu^2 + 2sp_0)[(p+q)^2 - \mu^2 + 2\bar{s}(p_0 + q_0)]}$$

approaches zero as $s_1 \to \pm\infty$, $s_2 \in \mathbb{C}^+$ *and as* $s_2 \to \infty$, *where* $s = s_1 + is_2$, *uniformly for all* $\mu \in \mathbb{R}$ *and* $q \in \mathbb{R}^3$. *The first limit is also uniform in* s_2 *and the second in* s_1.

Proof.

$$\mathscr{I} = \int_0^1 d\alpha \int_{\mathbb{R}^3} dp$$
$$\times \left[(p^2 - \mu^2 + 2sp_0)(1 - \alpha) + [(p + q)^2 - \mu^2 + 2\bar{s}(p_0 + q_0)]\alpha \right]^{-2}$$

$$= \int_0^1 d\alpha \int_{\mathbb{R}^3} dp$$
$$\times \left[(p + \alpha q)^2 - \beta q^2 - \mu^2 + 2s_1(p_0 + \alpha q_0) + is_2(\beta' p_0 - \alpha q_0) \right]^{-2}$$

$$= \int_0^1 d\alpha \int_{\mathbb{R}^3} dp \left[p^2 - \beta q^2 - \mu^2 + 2s_1 p_0 + is_2(\beta' p_0 + \beta q_0) \right]^{-2}$$

$$= \pi \int_0^1 d\alpha \int_{-\infty}^{\infty} dp_0 \left[p_0^2 + 2s_1 p_0 - \mu^2 - \beta q^2 + is_2(\beta' p_0 + \beta q_0) \right]^{-1}$$

$$= \frac{\pi}{s_1} \int_0^1 d\alpha \int_{-\infty}^{\infty} dt \left[t^2 + 2t - \frac{\mu^2}{s_1^2} - \frac{\beta}{s_1^2} q^2 + is_2 \left(\beta' \frac{t}{s_1} + \beta \frac{q_0}{s_1^2} \right) \right]^{-1}$$

$$= \frac{\pi}{s_1} \int_0^1 d\alpha \left(\int_{|t| < \epsilon} + \int_{|t| > \epsilon} \right) dt \dots,$$

where $\beta := \alpha(\alpha - 1)$ and $\beta' := 1 - 2\alpha$. The second integral clearly vanishes as $s_1 \to \pm\infty$, uniformly. In the first we may neglect the t^2 in the denominator and then carry out the t-integral:

$$\mathscr{I} = \frac{\pi}{s_1} \int_0^1 d\alpha \frac{1}{2 + is_2\beta'} \log \left[\frac{2\epsilon s_1^2 - \beta q^2 - \mu^2 + is_2(\beta q_0 + \beta' \epsilon s_1)}{-2\epsilon s_1^2 - \beta q^2 - \mu^2 + is_2(\beta q_0 - \beta' \epsilon s_1)} \right],$$

which approaches zero as $s_1 \to \pm\infty$, uniformly.

For the other limit we write

$$\mathscr{I} = -i\pi \int_0^1 d\alpha \int_{-\infty}^{\infty} dr \left[s_2(\beta' r + \beta q_0) + i(r^2 + 2s_1 r - \mu^2 - \beta q^2) \right]^{-1}$$

$$= -i\pi \int_0^1 \left(\int_{|r| < A} + \int_{|r| > A} \right) dr \dots.$$

The first vanishes uniformly as $s_2 \to \infty$. The second:

$$\int_0^1 d\alpha \int_{|r| > A} dr \, r^{-1}(ir + s_2\beta')^{-1} = \frac{1}{s_2} \int_0^1 d\alpha \int_{|t| > A/s_2} dt \, t^{-1} \frac{1}{it + \beta'},$$

which vanishes as $s_2 \to \infty$.   □

We now return to the proof of Lemma 6.2.2. Defining $v(p) := \int_{\mathbb{R}^3} dx \, |V(x)| e^{ip \cdot x}$, we find

$$I := \|\Upsilon_\gamma\|_2^2 = \int_{\mathbb{R}^3 \times \mathbb{R}^3} dx \, dy \, |V(x)| \, |V(y)|$$

$$\times \int_{\mathbb{R}^3 \times \mathbb{R}^3} dp \, dp' \frac{e^{i(p - p') \cdot (x - y)}}{(p^2 - \mu^2 + 2sp_0)(p'^2 - \mu^2 + 2\bar{s}p_0')}$$

$$= \int_{\mathbb{R}^3 \times \mathbb{R}^3} dp\, dp' \, \frac{|v(p - p')|^2}{(p^2 - \mu^2 + 2sp_0)(p'^2 - \mu^2 + 2\bar{s}p_0')}$$

$$= \int_{\mathbb{R}^3} dp'' \, |v(p'')|^2 \mathscr{J},$$

where $\mathscr{J}$ is the integral defined in Lemma 6.2.8. The assertion of Lemma 6.2.2 now follows from Lemma 6.2.8. $\square$

Proof of Lemma 6.2.5.

$$\|T\|_2^2 = \operatorname{tr} TT^\dagger = \int_{\mathbb{R}^3 \times \mathbb{R}^3} dx\, dy \, |V(x)| \, |V(y)| \, |B_\gamma^0(k, x - y)|^2$$

$$\leq \frac{1}{(2\pi)^4} \int_{\mathbb{R}^3 \times \mathbb{R}^3} dp\, dp' \, |v(p - p')|^2 \delta(k^2 - p^2) \delta(k^2 - p'^2)$$

$$= \frac{1}{(2\pi)^4} \int_{\mathbb{R}^3 \times \mathbb{R}^3} dp\, dp' \, |v(p')|^2 \delta(k^2 - p^2) \delta(p'^2 + p \cdot p')$$

$$= \frac{1}{(2\pi)^4} \int_{S^2 \times S^2} d\hat{p}\, d\hat{p}' \, |k|^2 \hat{p} \cdot \hat{p}' \Theta(\hat{p} \cdot \hat{p}') |v(2|k|\hat{p} \cdot \hat{p}\hat{p}')|^2$$

$$= \frac{1}{(2\pi)^4} \int_{S^2} d\hat{p}' \int_0^{2\pi} d\phi \int_0^{|k|} du\, u |v(2u\hat{p}')|^2$$

$$\leq \frac{1}{2\pi^2} \int_0^\infty du\, u [V_1(2u)]^2 < C.$$

Here $\hat{p} = p/|p|$. This proves the first part of the lemma.

For the second part define

$$C_\gamma(k, x, y) := \int_{\mathbb{R}^3} dz\, B_\gamma^0(k, x - z) |V(z)| B_\gamma^0(k, z - y)$$

$$= -\frac{1}{(2\pi)^4} \int_{\mathbb{R}^3 \times \mathbb{R}^3} dp\, dp' \, e^{i(p \cdot x - p' \cdot y)} \Theta[\gamma \cdot (p - k)]$$

$$\times \Theta[\gamma \cdot (p' - k)] \delta(p^2 - k^2) \delta(p'^2 - k^2) v(p - p').$$

Then we have

$$I := \|TT^\dagger\|_2^2 = \int_{\mathbb{R}^3 \times \mathbb{R}^3} dx\, dy \, |V(x)| \, |V(y)| \, |C_\gamma(k, x, y)|^2$$

$$= \frac{1}{(2\pi)^8} \int dp\, dp'\, dq\, dq' \, \Theta[\gamma \cdot (p - k)] \Theta[\gamma \cdot (p' - k)] \Theta[\gamma \cdot (q - k)]$$

$$\times \Theta[\gamma \cdot (q' - k)] \delta(p^2 - k^2) \delta(p'^2 - k^2) \delta(q^2 - k^2) \delta(q'^2 - k^2)$$

$$\times v(p - q) v(q' - p') v(p - p') v(q - q').$$

The variables of integration p and q' are now shifted according to $p = p'' + p'$, $q' = q'' + q$, and one obtains

$$I \leq \frac{1}{(2\pi)^8} \int dp'\, dp''\, dq\, dq'' \, \delta(p''^2 + 2p' \cdot p'') \delta(q''^2 + 2q \cdot q'') \delta(p'^2 - k^2)$$

$$\times \delta(q^2 - k^2) |v(p'')| \, |v(q'')| \, |v(p' + p'' - q)| \, |v(q + q'' - p')|$$

$$\leq \frac{|k|^4}{(2\pi)^8} \int d\hat{p}' \, d\hat{p}'' \, d\hat{q} \, d\hat{q}'' \, |v(-2|k|\hat{p}''\hat{p}'' \cdot \hat{p}')| \, |v(2|k|\hat{q}''\hat{q} \cdot \hat{q}'')| |\hat{p}' \cdot \hat{p}''|$$
$$\times |\hat{q} \cdot \hat{q}''| \, |v[|k|(\hat{p}' - \hat{q} - 2\hat{p}''\hat{p}' \cdot \hat{p}'')]| \, |v[|k|(\hat{q} - \hat{p}' - 2\hat{q}''\hat{q} \cdot \hat{q}'')]|.$$

Using spherical polar coordinates for the $\hat{q}''$ and $\hat{p}''$ integration over $\mathbf{S}^2$, with $\hat{q}$ and $\hat{p}'$, respectively, as the z-axes, one gets

$$I \leq \frac{|k|^4}{(2\pi)^8} \int_{\mathbf{S}^2 \times \mathbf{S}^2} d\hat{q} \, d\hat{p}' \int_0^{2\pi} d\phi_1 \int_0^{2\pi} d\phi_2 \int_{-1}^{1} du_1 \int_{-1}^{1} du_2 |v(-2|k|u_1\hat{p}'')|$$
$$\times |v(2|k|u_2\hat{q}'')| \, |v[|k|(\hat{p}' - \hat{q} - 2u_1\hat{p}'')]| \, |v[|k|(\hat{q} - \hat{p}' - 2u_2\hat{q}'')]||u_1u_2|$$
$$= \frac{1}{(2\pi)^8} \int d\hat{q} \, d\hat{p}' \int_0^{|k|} dl_1 \, l_1 \int_0^{|k|} dl_2 \, l_2 \int_0^{2\pi} d\phi_1 \int_0^{2\pi} d\phi_2 |v(2l_1\hat{p}'')|$$
$$\times |v(-2l_2\hat{q}'')| \, |v[2l_1\hat{p}'' + |k|(\hat{p}' - \hat{q})]| \, |v[2l_2\hat{q}'' + |k|(\hat{q} - \hat{p}')]|.$$

By the assumption that $|v(p)| \leq V_1(|p|)$, $\int_0^\infty dt \, tV_1(t) < \infty$, one may extend the l-integrals to ∞. They are then each split into two pieces, from 0 to A, and from A to ∞. Since v is uniformly bounded the last part may be made smaller than any ϵ by choosing A sufficiently large. In the first part we have

$$\lim_{|k|\to\infty} \int_{\mathbf{S}^2 \times \mathbf{S}^2} d\hat{q} \, d\hat{p}' \, |v([|k|(\hat{p}' - \hat{q})]|^2$$
$$= \lim_{|k|\to\infty} \int_{\mathbf{S}^2 \times \mathbf{S}^2} d\hat{q} \, d\hat{p}' \int_{\mathbb{R}^3 \times \mathbb{R}^3} dx \, dy \, |V(x)| \, |V(y)| e^{i|k|(x-y)\cdot(\hat{p}'-\hat{q})} = 0$$

by the Riemann-Lebesgue lemma.   □

Proof of Lemma 6.2.7.

Lemma 6.2.9. *Let $L(k)$ be the operator family defined by the integral kernels (1.13) and let V be in the Rollnik class. Then $\exists C$ such that for all $k \in \mathbb{R}$, $\|L(k)\|_2 < C$, and furthermore,*

$$\lim_{k\to\pm\infty} \|L(k)L^\dagger(k)\|_2 = 0.$$

Here $\| \cdot \|_2$ is the Hilbert-Schmidt norm.

Proof.

$$\mathrm{tr} LL^\dagger = \frac{1}{(4\pi)^2} \int_{\mathbb{R}^3 \times \mathbb{R}^3} dx \, dy \, |V(x)| \, |V(y)| \frac{1}{|x-y|^2} < C$$

by assumption. The second part of the lemma is a theorem due to Zemach and Klein [ZK58].   □

We proceed with the proof of Lemma 6.2.7. Define

$$L' := (\mathbb{1} - L)^{-1}L, \ L'' := (\mathbb{1} - L)^{-1} = \mathbb{1} + L', \ T := |V|^{\frac{1}{2}} B_\gamma^0 V^{\frac{1}{2}}.$$

We have by (7.2)

$$D_\gamma(\mu, s) = \Delta(|k|) \det(\mathbb{1} - B_\gamma^+ V)e^{\mathrm{tr} B_\gamma^0 V},$$

and hence

$$\arg D_\gamma(\mu, s) = \arg \Delta(|k|) + \arg\left[\det(\mathbb{1} - B_\gamma^+ V)e^{\operatorname{tr}B_\gamma^0 V}\right],$$

where $\Delta(|k|)$ is the Fredholm determinant of the Lippmann-Schwinger equation, and

$$B_\gamma^+ V = (\mathbb{1} - G_0^+ V)^{-1} B_\gamma^0 V.$$

The function B_γ^0 depends on s and μ^2, and $|k| = \sqrt{s^2 + \mu^2}$ so that $|k| \to +\infty$ as $s \to +\infty$. Now,

$$\operatorname{tr}B_\gamma^+ V = \operatorname{tr}T + \operatorname{tr}LT + \operatorname{tr}L'LT,$$

and

$$
\begin{aligned}
|\operatorname{tr}L'LT| &= |\operatorname{tr}(\mathbb{1} - L^2)^{-1}(\mathbb{1} + L)L^2 T| \\
&\le (1 - \|L^2\|)^{-1}\|\mathbb{1} + L\|\,\|T\|_2\|L^2\|_2 \longrightarrow 0
\end{aligned}
$$

as $\mu^2 + s^2 \to \infty$, by (1.97), (1.98), and Lemma 6.2.5. Hence

$$\operatorname{tr}B_\gamma^+ V = \operatorname{tr}T + \operatorname{tr}LT + o(1). \tag{6.29}$$

For $n \ge 1$

$$
\begin{aligned}
|\operatorname{tr}(B_\gamma^+ V)^{2n}| &= |\operatorname{tr}(L''T)^{2n}| \le \|(L''T)^n\|_2^2 = \operatorname{tr}(L''T)^n(T^\dagger L''^\dagger)^n \\
&= \operatorname{tr}(L''^\dagger L'')(TL'')^{n-1}(TT^\dagger)(L''^\dagger T^\dagger)^{n-1} \\
&\le \|(L''^\dagger L'')(TL'')^{n-1}\|_2\|(TT^\dagger)(L''^\dagger T^\dagger)^{n-1}\|_2 \\
&\le \|(L''^\dagger L'')\|_2\|TT^\dagger\|_2\|(TL'')^{n-1}\|_2 \\
&\le \ldots \le \|L''^\dagger L''\|_2^n\|TT^\dagger\|_2^n.
\end{aligned}
$$

Similarly,

$$
\begin{aligned}
|\operatorname{tr}(B_\gamma^+ V)^{2n+1}|^2 &= |\operatorname{tr}(L''T)^{2n+1}|^2 \le \|(L''T)^n\|_2^2\|(L''T)^{n+1}\|_2^2 \\
&\le \|L''^\dagger L''\|_2^{2n+1}\|TT^\dagger\|_2^{2n+1},
\end{aligned}
$$

and therefore for sufficiently large $|k|$ the series in the exponential expression for $\det(\mathbb{1} - B_\gamma^+ V)$ converges and for $n \ge 2$

$$\operatorname{tr}(B_\gamma^+ V)^n = o(1) \tag{6.30}$$

by (1.97), (1.98), and Lemma 6.2.5. It follows from equations (6.29) and (6.30) that

$$\lim_{\mu^2 + s^2 \to \infty} \det(\mathbb{1} - B_\gamma^+ V)e^{\operatorname{tr}B_\gamma^0 V} = \lim_{\mu^2 + s^2 \to \infty} e^{-\operatorname{tr}LT - \frac{1}{2}\operatorname{tr}T^2}. \tag{6.31}$$

But

$$\overline{\operatorname{tr}T^2} = \operatorname{tr}V\widetilde{B_\gamma^0}V\widetilde{B_\gamma^0} = \operatorname{tr}V B_\gamma^0 V B_\gamma^0 = \operatorname{tr}T^2,$$

since $\widetilde{B_\gamma^0} = -B_\gamma^0$. Hence the factor $\exp(-\frac{1}{2}\operatorname{tr}T^2)$ does not contribute to $\arg D_\gamma$. Finally we examine $\operatorname{tr}LT$:

$$
\begin{aligned}
\mathrm{tr}LT &= -\frac{i|k|}{4(2\pi)^3} \int_{\mathbb{R}^3 \times \mathbb{R}^3} dx\, dy\, V(x)V(y) \frac{e^{i|k|\,|x-y|}}{|x-y|} \\
&\quad \times \int_{S^2} d\theta\, e^{i|k|\theta \cdot (x-y)} \Theta(\gamma \cdot \theta|k| - s) \\
&= -\frac{i|k|}{4(2\pi)^3} \int_{\mathbb{R}^3 \times \mathbb{R}^3} dx\, dy\, |y|^{-1} V(x)V(y+x) \\
&\quad \times \int_{S^2} d\theta\, e^{i|k||y|(1-\theta\cdot\hat{y})} \Theta(\gamma \cdot \theta|k| - s) \\
&= \frac{1}{4(2\pi)^3} \int_{\mathbb{R}^3 \times \mathbb{R}^3} dx\, dy\, |y|^{-1} V(x)V(x+y)\Theta\left(\gamma \cdot \theta - \frac{s}{|k|}\right) + o(1).
\end{aligned}
$$

This term is real and thus makes no contribution to $\arg D_y$. Thus, as $\mu^2 + s^2 \to \infty$, $\Im \mathrm{tr} LT = o(1)$. The lemma follows because $\arg \Delta(k) = o(1)$ as $k \to +\infty$.   □

6.3 Exceptional Points

The exceptional points for the integral equation (6.16) are those pairs $\mu \in \mathbb{R}_+$, $s \in \mathbb{C}^+ \cup \mathbb{R}$ for which the modified Fredholm determinant of (6.20) vanishes,

$$D_\gamma(\mu, s) = 0.$$

Since D_y is an analytic function of s that is holomorphic in $\mathbb{C}^+$, each of its zeros there is of finite multiplicity. Whether this is generally true also on the real axis is unknown. If $\exists \epsilon > 0$ such that (1.80) holds, then D_y is analytic also on the real axis and thus its zeros there, if any, are of finite multiplicity too. The reality of the potential together with the symmetry (6.8) implies that whenever, for a given value of μ, s is an exceptional point, then so is $-\bar{s}$ for the same value of μ. Furthermore, the multiplicities of the two corresponding zeros of D_y are equal.

Corollary 6.2.3 implies that there must exist a semi-circle in the upper half-plane, centered at the origin, outside of which there are no exceptional points. Therefore, if their total number is infinite, they have to accumulate somewhere on the real axis. The multiplicity of the zero of D_y at such a point of accumulation is necessarily infinite. We may thus conclude the following.

Lemma 6.3.1. *If the multiplicity of each zero of $D_y(\mu, s)$ as a function of s on the real axis is finite then the total number of exceptional points is finite. In particular, if $\exists \epsilon > 0$ such that (1.80) holds then that number is finite.*

As $\mu \to 0$ we have (6.24). Therefore for $\mu = 0$ the exceptional points of (6.16) coincide with those of (1.3); for $s \in \mathbb{C}^+$ they are the bound states in the sense that s^2 is an L^2-eigenvalue of the Schrödinger equation. Since $F_y(\mu, s, x)$ is a continuous function of μ, the functions $s(\mu)$ defined by $D_y(\mu, s(\mu)) = 0$ describe *zero-trajectories* in $\mathbb{C}^+$. Each bound-state point on the positive imaginary axis is the starting point of such a trajectory for $\mu = 0$. Consider one of these. As μ increases from zero, the corresponding zero-trajectory must remain on the imaginary axis until it collides with another coming from the opposite direction.

The fact that $D_\gamma(\mu, s)$ is an analytic function of s and continuous as a function of μ implies a "conservation law" of zeros: as μ changes, the decrease of the number of zeros (counting their multiplicities) in any region of analyticity equals the number of zeros that cross the boundary of that region. What is more, two symmetrically (with respect to the imaginary axis) situated zeros that collide on the imaginary axis must necessarily "bounce off" in opposite directions, and *vice versa*. Therefore, a double zero moving along the imaginary axis cannot split into two simple zeros that become symmetric trajectories in the first and second quadrant of the complex plane.

The zero trajectories of D_γ are at the same time the pole trajectories of φ_γ. Since the solutions φ_γ approach the scattering solutions as $\mu \to 0$, and the latter always have *simple* poles at the exceptional points in $\mathbb{C}^+$ we make the following

Conjecture 6.3.2. *If* $V \in \mathcal{V}$, *then the functions* $\varphi_\gamma(\mu, s, x)$ *and* $\xi_\gamma(k_\perp, s, x)$ *have simple poles at the exceptional points in* $\mathbb{C}^+$.

Remark. The reason why the scattering solution has simple poles only is that the resolvent of a self-adjoint operator necessarily has simple poles at the eigenvalues. The Faddeev Green's function is the kernel of the resolvent of $H = V - \Delta$ on a weighted L^2-space, where H is not self-adjoint. Nevertheless, it may be expected that the connection of the pole trajectories to their simple-pole end-points at $\mu = 0$ makes the poles for $\mu > 0$ simple also, except possibly when two trajectories cross.

An important question that remained open for about fifteen years is whether there are ever any *real exceptional points*. These may be zero-trajectories that cross or osculate the real axis, or they may not be connected with zero-trajectories in $\mathbb{C}^+$ at all. An answer was recently given by R. G. Novikov and G. M. Henkin and by R. Lavine and A. Nachman:

Theorem 6.3.3. (NHLN) *Suppose that* $V \in \mathcal{V}$. *If* $D_\gamma(\mu, s) = 0$ *for any* $\mu \geq 0$ *and* $s \in \mathbb{C}^+$, *then there exists a pair* $\mu > 0$ *and* $s \in \mathbb{R}$ *such that* $D_\gamma(\mu, s) = 0$. *Furthermore, there exists a* $\mu_0 \geq 0$ *such that for all* $\mu \geq \mu_0$ $D_\gamma(\mu, s)$ *has no zeros for* $s \in \mathbb{C}^+$.

Proof. By the argument principle the number $n(\mu)$ of zeros of $D_\gamma(\mu, s)$ regarded as a function of $s \in \mathbb{C}^+$ is such that

$$2\pi n(\mu) = \lim_{s \to \infty} \arg D_\gamma(\mu, s) - \arg D_\gamma(\mu, 0+)$$
$$+ \arg D_\gamma(\mu, 0-) - \lim_{s \to -\infty} \arg D_\gamma(\mu, s)$$

because of the first part of Corollary 6.2.3. Because of the symmetry (6.22) we may choose $\arg D_\gamma(\mu, -s) = -\arg D_\gamma(\mu, s)$ (at the expense of a discontinuity at $s = 0$) and thus obtain by the use of Lemma 6.2.7

$$\pi n(\mu) = -\arg D_\gamma(\mu, 0+)$$

if we choose the phase of D_γ to be zero at $s \to \infty$. A second use of Lemma 6.2.7, this time letting $\mu \to \infty$, implies that $\exists \mu_0 < \infty$ such that for $\mu > \mu_0$, $n = 0$. By Corollary 6.2.3 a zero of D_γ cannot, as a function of μ, move to infinity. Hence every zero that for some μ was in $\mathbb{C}^+$ must cross the real axis as $\mu \to \infty$. □

In particular, this theorem implies that whenever there are bound states there must be real exceptional points. However, it does not give any information about where the zero-trajectories that begin at the bound states cross the real axis, nor does it give an upper limit on the number of real exceptional points. It would not violate the theorem if all the real exceptional points occurred at the origin, $s = 0$. The question whether this is, perhaps, always the case is open. And so is the question whether there are other exceptional points that have no relation to bound states, or whether their number is necessarily finite.

The significance of the exceptional points, physical or otherwise, is quite unclear. In contrast to the exceptional points in $\mathbb{C}^+$ of the Lippmann-Schwinger equation those of equation (6.16) do not necessarily indicate the existence of square-integrable solutions. If s is a *real* exceptional point then there exists a solution of the Schrödinger equation that solves the homogeneous version of (6.16) and hence behaves asymptotically for large $|x|$ like the Green's function F_γ, which is given by (6.14) or (6.15). For such a solution there exists a double cone outside of which there are only outgoing (or incoming, depending on the sign of s) spherical waves at infinity, and in half of which the function decreases faster than $1/r$. A meaningful physical interpretation of such a solution is lacking.

In order to see the asymptotic behavior of the solutions of (6.16) at exceptional points in $\mathbb{C}^+$ we consider the equation (6.19). The asymptotic form of the function $\mathscr{F}_\gamma$ for large $r = |x|$ and $s \in \mathbb{C}^+$ is readily obtained from (6.4) as

$$\mathscr{F}_\gamma(k_\perp, s, x) = \frac{i\mu}{4\pi s |x_\perp|} J_1(\mu|x_\perp|)e^{ik_\perp \cdot x_\perp} + o\left(\frac{1}{|x_0|}\right).$$

Therefore, if $s \in \mathbb{C}^+$ is an exceptional point of (6.16) then the asymptotic form of a solution of its homogeneous version is given by

$$\xi_\gamma(k_\perp, s, x) = e^{-ik_\perp \cdot x_\perp}\left[\mathscr{Y}_\gamma(\mu, s, x_\perp) + o(1/|x_0|)\right], \tag{6.32}$$

where

$$\mathscr{Y}_\gamma(\mu, s, x_\perp) := \frac{i\mu}{4\pi s}\int_{\mathbb{R}^3} dy \, \frac{V(y)}{|x_\perp - y_\perp|} J_1(\mu|x_\perp - y_\perp|)\xi_\gamma(k_\perp, s, y)e^{ik_\perp \cdot y_\perp}. \tag{6.33}$$

The functions $\mathscr{Y}_\gamma$ are the analogues of the characters of the bound-state eigenfunctions defined in Section 1.2.

6.4 Large-r Asymptotics and Scattering

The large-distance asymptotics of the solution of the integral equation (6.16) for real s is readily obtained from that equation together with (6.14). One obtains

$$\varphi_\gamma^+(k_\perp, s, x) = e^{ik \cdot x} + r^{-1}e^{i|k|r}\Theta(sr - |k|x_0)\widehat{h}_\gamma(|k|, \hat{x}, \hat{k}) \tag{6.34}$$
$$+ r^{-1}e^{-i|k|r}\Theta(-sr - |k|x_0)\widehat{h}_\gamma(|k|, -\hat{x}, \hat{k}) + o(r^{-1}),$$

where Θ is the Heaviside function and

$$\widehat{h}_\gamma(|k|, \hat{k}', \hat{k}) := -\frac{1}{4\pi}\int_{\mathbb{R}^3} dx \, e^{-ik' \cdot x}V(x)\varphi_\gamma^+(k_\perp, s, x) \tag{6.35}$$

with $k = |k|\hat{k}$, $k' = |k|\hat{k}'$. It is sometimes useful to write $\hat{h}_\gamma$ as a function of different variables as follows:

$$\hat{h}_\gamma(|k|, \hat{k}', \hat{k}) = h_\gamma(k'_\perp, s', k_\perp, s), \tag{6.36}$$

with the understanding that $k'^2_\perp + s'^2 = k^2_\perp + s^2 = |k|^2$.

The function $\hat{h}_\gamma$ plays a role that is similar to that of the scattering amplitude for the scattering solution, and its integral representation resembles that given in (1.62). Nevertheless, this function has no direct relation to scattering because φ_γ does not describe scattering. It follows from the reality of the potential and the symmetry (6.8) that $\hat{h}_\gamma$ has the symmetry

$$\hat{h}_\gamma(|k|, -\theta', -\theta) = \overline{\hat{h}_\gamma(|k|, \theta', \theta)}. \tag{6.37}$$

In the notation of the function h_γ (6.37) reads

$$\overline{h_\gamma(k'_\perp, s', k_\perp, s)} = h_\gamma(-k'_\perp, -s', -k_\perp, -s). \tag{6.38}$$

The symmetry (6.9) leads to

$$h_\gamma(k'_\perp, s', k_\perp, s) = h_{-\gamma}(-k_\perp, s, -k'_\perp, s'), \tag{6.39}$$

or

$$\hat{h}_\gamma(|k|, \theta', \theta) = \hat{h}_{-\gamma}(|k|, -\theta, -\theta'). \tag{6.40}$$

The analyticity properties of $\varphi_\gamma(k_\perp, s, x)$ as a function of s imply that the function $h_\gamma(k'_\perp, s, k_\perp, s)$ has an analytic continutation into $\mathbb{C}^+$ with poles at the exceptional points of (6.16).

Definition 6.4.1. The function $f_\gamma(|k|, \hat{k}', \hat{k})$, $\mathbb{R}_+ \times S^2 \times S^2 \times S^2 \mapsto \mathbb{C}$, which will also be denoted by $\hat{f}_\gamma(k'_\perp, s', k_\perp, s)$, where $s = \gamma \cdot \hat{k}|k|$, $s' = \gamma \cdot \hat{k}'|k|$, $k_\perp = \hat{k}|k| - \gamma s$, $k'_\perp = \hat{k}'|k| - \gamma s'$, is in the class $\mathscr{T}$ if the function $\hat{f}_\gamma(k'_\perp, s, k_\perp, s)$ has the following properties for each $\gamma \in S^2$, $k_\perp, k'_\perp \in \mathbb{R}^2$, $|k_\perp| = |k'_\perp|$:

(i) For almost all $|k| \in \mathbb{R}_+$ it is a continuous function of $\hat{k}$ and $\hat{k}'$.

(ii) It is the boundary value of an analytic function of s that is holomorphic in $\mathbb{C}^+$.

(iii) Its limit as $|s| \to \infty$ in $\mathbb{C}^+$ exists, is independent of γ, and is a function of $k_\perp - k'_\perp$ only:

$$\lim_{|s| \to \infty} \hat{f}_\gamma(k'_\perp, s, k_\perp, s) = C(k_\perp - k'_\perp).$$

We define

$$H_\gamma(k'_\perp, k_\perp, s) := D_\gamma(|k_\perp|, s) h_\gamma(k'_\perp, s, k_\perp, s),$$

where D_γ is the modified Fredholm determinant of (6.16). As a result of (6.23) together with (6.22),(6.38), and (6.39) one finds that H_γ enjoys the following symmetry relations for $s \in \mathbb{R}$:

$$H_\gamma(k'_\perp, k_\perp, -s) = \overline{H_\gamma(-k'_\perp, -k_\perp, s)}, \tag{6.41}$$

$$H_\gamma(k'_\perp, k_\perp, s) = \overline{H_{-\gamma}(k_\perp, k'_\perp, s)}. \tag{6.42}$$

We then have the following result.

Lemma 6.4.2. *If $V \in \mathcal{V}$ then the function $H_\gamma(k'_\perp, k_\perp, s)$ is in $\mathcal{T}$ and the limit C is given by*

$$C(k_\perp - k'_\perp) = -\frac{1}{4\pi} \int_{\mathbb{R}^3} dx \, V(x) e^{ix \cdot (k_\perp - k'_\perp)}. \tag{6.43}$$

The last part of the lemma follows from (6.35), Corollary 6.2.3, and Lemma 6.2.4.

We now want to connect the Faddeev solution to the scattering solution. According to (6.10) the latter is a special case of the former. By that equation and (6.2) we have for $|k| \in \mathbb{R}_+$

$$\psi^\pm(|k|, \theta, x) = \widehat{\varphi}_\theta^\pm(|k|\theta, x) = \widehat{\varphi}_{\pm\theta}^+(|k|\theta, x). \tag{6.44}$$

Similarly for the scattering amplitude,

$$\widehat{h}_\theta(|k|, \theta', \theta) = A(|k|, \theta', \theta), \tag{6.45}$$

and

$$\widehat{h}_{-\theta}(|k|, \theta', \theta) = \overline{A}(|k|, \theta, \theta'), \tag{6.46}$$

by (1.70).

Lemma 1.5.1 assures us that the solution φ_γ of the Schrödinger equation can be expressed as a superposition of scattering solutions. For each $|k| \in \mathbb{R}_+$ there must exist a distribution kernel Δ_γ such that

$$\widehat{\varphi}_\gamma^+(|k|\theta, x) = \int_{S^2} d\theta' \, \Delta_\gamma(|k|, \theta', \theta) \psi^+(|k|, \theta', x), \tag{6.47}$$

and the lemma allows us to read off Δ_γ from the asymptotic form (6.34). The result obtained by using (1.59) is that

$$\Delta_\gamma(|k|, \theta', \theta) = \delta(\theta', \theta) + \frac{|k|}{2\pi i} q_\gamma(|k|, \theta', \theta), \tag{6.48}$$

where

$$q_\gamma(|k|, \theta', \theta) = \widehat{h}_\gamma(|k|, \theta', \theta) \Theta[\gamma \cdot (\theta' - \theta)]. \tag{6.49}$$

Furthermore Corollary 1.5.2 and the symmetry (6.37) lead to the equation

$$\overline{\Delta}_\gamma(|k|, \theta, \theta') = \int_{S^2} d\theta'' \, S(|k|, -\theta, \theta'') \Delta_\gamma(|k|, \theta'', -\theta'). \tag{6.50}$$

Use of the form (6.48) of Δ_γ and of the S matrix in terms of the scattering amplitude together with the symmetry (6.37) allows us to write this equation in the form

$$\widehat{h}_\gamma(|k|, \theta, \theta') = A(|k|, \theta, \theta')$$
$$+ \frac{|k|}{2\pi i} \int_{S^2} d\theta'' \, A(|k|, \theta, \theta'') \widehat{h}_\gamma(|k|, \theta'', \theta') \Theta[\gamma \cdot (\theta'' - \theta')]. \tag{6.51}$$

It follows from this, together with the unitarity equation (1.72), that

$$\Delta_\theta(|k|, \theta', \theta) = \delta(\theta', \theta), \tag{6.52}$$

and

$$\Lambda_{-\theta}(|k|, \theta', \theta) = \overline{S(|k|, \theta', \theta)}. \tag{6.53}$$

Straight-forward computation, using (6.48) and (6.49) gives the following result:

$$
\begin{aligned}
\Lambda_\gamma(|k|, \theta', \theta) &:= \int_{S^2} d\theta'' \, \overline{\tilde{\Lambda}_{-\gamma}(|k|, \theta', \theta'')} \Lambda_\gamma(|k|, \theta'', \theta) - \delta(\theta', \theta) \\
&= \frac{|k|}{2\pi i} \xi_\gamma(|k|, \theta', \theta),
\end{aligned}
$$

where the tilde denotes the transpose of the kernel and

$$
\begin{aligned}
\xi_\gamma(|k|, \theta', \theta) &= \Theta[\gamma \cdot (\theta' - \theta)] \left[\hat{h}_\gamma(|k|, \theta', \theta) - \overline{\hat{h}_{-\gamma}(|k|, \theta, \theta')} \right. \\
&\quad \left. - \frac{|k|}{2\pi i} \int_{\gamma \cdot \theta < \gamma \cdot \theta'' < \gamma \cdot \theta'} d\theta'' \, \overline{\hat{h}_{-\gamma}(|k|, \theta'', \theta')} \hat{h}_\gamma(|k|, \theta'', \theta) \right].
\end{aligned}
$$

On the other hand, (6.50) together with (1.72) and (1.70) leads to

$$\Lambda_\gamma(|k|, \theta', \theta) = -\overline{\Lambda_\gamma(|k|, -\theta', -\theta)},$$

which implies that $\xi_\gamma(|k|, \theta', \theta) = \overline{\xi_\gamma(|k|, -\theta', -\theta)}$. But since $\xi_\gamma(|k|, \theta', \theta) = 0$ for $\gamma \cdot \theta > \gamma \cdot \theta'$ and by Lemma 6.4.2 it is bounded, it follows that $\xi_\gamma = 0$. Consequently, $\Lambda_\gamma = 0$. Regarding $\Lambda_\gamma(|k|, \theta', \theta)$ as the integral kernels of the operator family $\Lambda_\gamma(|k|)$, one may write this result in the simple form

$$\overline{\tilde{\Lambda}_{-\gamma}(|k|)} \Lambda_\gamma(|k|) = \mathbb{1}, \tag{6.54}$$

which says that $\overline{\tilde{\Lambda}_{-\gamma}}$ is the left inverse of Λ_γ. Suppose that $\Lambda'_\gamma = \mathbb{1} + \lambda_\gamma$ is a right inverse; then the kernel of the operator λ_γ must satisfy the quasi-Volterra equation

$$\lambda_\gamma(|k|, \theta, \theta') + \Theta[\gamma \cdot (\theta - \theta')] \hat{h}_\gamma(|k|, \theta, \theta')$$

$$+ \frac{|k|}{2\pi i} \int_{S^2} d\theta'' \, \Theta[\gamma \cdot (\theta - \theta'')] \hat{h}_\gamma(|k|, \theta, \theta'') \lambda_\gamma(|k|, \theta'', \theta') = 0,$$

which always has a unique solution. The left inverse of Λ_γ similarly satisfies a quasi-Volterra equation and therefore is unique. Therefore Λ_γ has an inverse and it is given by $\overline{\tilde{\Lambda}_{-\gamma}}$:

$$[\Lambda_\gamma(|k|)]^{-1} = \overline{\tilde{\Lambda}_{-\gamma}(|k|)}. \tag{6.55}$$

Inserting the form (6.48) in (6.54) results in the equation $q_\gamma - \overline{\tilde{q}}_{-\gamma} - \frac{|k|}{2\pi i} q_\gamma \overline{\tilde{q}}_{-\gamma} = 0$. Notice, however, that (6.40), (6.37), and (6.49) together imply that

$$\overline{\tilde{q}}_{-\gamma} = q_\gamma. \tag{6.56}$$

Therefore (6.55) implies that q_γ is nilpotent:

$$q_\gamma^2 = 0. \tag{6.57}$$

Written out explicitly, this equation reads for $\gamma \cdot (\theta - \theta') > 0$,

$$\int_{\gamma\cdot\theta'<\gamma\cdot\theta''<\gamma\cdot\theta} d\theta'' \,\widehat{h}_\gamma(|k|,\theta,\theta'')\widehat{h}_\gamma(|k|,\theta'',\theta') = 0. \tag{6.58}$$

As a result equation (6.50) may be rewritten in the form

$$\bar{S} = Q\Delta_\gamma Q\widetilde{\Delta}_{-\gamma}, \tag{6.59}$$

in which Q is the same operator used earlier and defined above Definition 1.5.13. Equation (6.59), together with (6.40), written out in detail, reads as follows:

$$\overline{A(|k|,\theta',\theta)} = \widehat{h}_\gamma(|k|,\theta,\theta')$$

$$+\frac{|k|}{2\pi i} \int_{S^2} d\theta'' \,\Theta\,[\gamma\cdot(\theta-\theta'')]\Theta\,[\gamma\cdot(\theta'-\theta'')]\widehat{h}_\gamma(|k|,\theta,\theta'')\widehat{h}_\gamma(|k|,\theta'',\theta'). \tag{6.60}$$

The factorization (6.59) of the S matrix clearly differs from that defined by the Jost function in (2.77). It is characterized by two properties of the function Δ_γ: (a) (6.48) shows that Δ_γ differs from the identity by an operator that is lower triangular with respect to the direction γ, and (b) (6.49) together with Lemma 6.4.2 shows that when $k\cdot\gamma \to k'\cdot\gamma = s$ then $\Delta_\gamma - \mathbb{1}$ approaches a function that is the boundary value on the real axis of an analytic function of s meromorphic in $\mathbb{C}^+$.

Equation (6.59) allows us to make a direct connection between the determinant of the S matrix and the function h_γ. Using (6.48) and (6.49) leads to $\det S = \exp\frac{|k|}{2\pi}\mathrm{tr}[Q\bar{q}_\gamma Q + \widetilde{q}_{-\gamma}]$ because the lower triangularity of q_γ implies that $\mathrm{tr}\,q_\gamma^n = 0$ for all $n > 1$. Furthermore, $Qq_\gamma Q + \widetilde{q}_{-\gamma} = \bar{h}_\gamma$; therefore,

$$\det S(|k|) = \exp\left[\frac{i|k|}{2\pi} \int_{S^2} d\theta\,\widehat{h}_\gamma(|k|,\theta,\theta)\right]. \tag{6.61}$$

Equation (6.59) must hold for arbitrary directions γ. Together with (6.50) this implies that for any two $\gamma,\gamma' \in S^2$ the following equation must hold

$$\widetilde{\Delta}_{\gamma'}\overline{\Delta}_\gamma = Q\widetilde{\Delta}_{\gamma'}\Delta_\gamma Q. \tag{6.62}$$

Using (6.48), (6.49), and (6.37), this equation may be written explicitly in the form

$$\widehat{h}_\gamma(|k|,\theta',\theta) = \overline{\widehat{h}_{\gamma'}(|k|,\theta,\theta')} \tag{6.63}$$

$$+\frac{|k|}{2\pi i} \int_{S^2} d\theta''\,\widehat{h}_\gamma(|k|,\theta'',\theta)\overline{\widehat{h}_{\gamma'}(|k|,\theta'',\theta')}\Big[\Theta\,[\gamma'\cdot(\theta''-\theta')] - \Theta\,[\gamma\cdot(\theta-\theta'')]\Big].$$

In the special case in which $\gamma' = -\theta'$, (6.46) shows that this equation goes over into (6.51).

6.5 Notes

6.1 The Green's function given by (6.1) was introduced by Faddeev in [Fa65 and 66] for the specific purpose of eventually solving the three-dimensional inverse scattering problem. The split-ups (6.4) and (6.5) were given in [Fa65] without proofs.

6.2 The results of this section were first obtained by [Fa65, 71, and 74] and by [Ne74]. The statements of Lemma 6.2.2 and of its Corollary 6.2.3 are new refinements, and Lemma 6.2.7 is also new.

6.3 The contents of this section are partly new. Real exceptional points were discussed in [Ne87]. Theorem NHLN (though not exactly with the same hypotheses as stated here) was proved independently by Novikov and Henkin in [NH86, 87b, see also HN88] and by Lavine and Nachman in [LN86a, 86b, see also Na86]. The details of the proof by Lavine and Nachman have not been published as of this writing, and those of the proof by Novikov and Henkin have become available in English only after the draft of this book was finished. The proof presented here, based on Corollary 6.2.3 and Lemma 6.2.7, is independent of those given by these earlier authors, though I have benefited from conversations with A.Nachman about the ideas of his proof with Lavine. I have also benefited from conversations with R.G.Novikov and G.M.Henkin about the exceptional points.

6.4 Most of the contents of this section can be found in [Fa71 and 74] and in [Ne 74a and 74b]. However, equation (6.43) was given first in [Ne85 and NH87a].

7. The Inverse Problem

7.1 The Scattering Amplitude as Input

The first step in the solution of the inverse scattering problem by means of the Faddeev solution of the Schrödinger equation is to find the kernel h_γ from the given scattering amplitude. This can be done very simply by solving the integral equation (6.51). It should be noted that (6.51) is a Fredholm equation in which the value of $|k|$ is fixed. It makes sense only for real s. Let us call its Fredholm determinant $d_\gamma(\mu, s)$,

$$d_\gamma(\mu, s) := \det[1 - A_\gamma(k)], \tag{7.1}$$

where A_γ is the operator whose kernel is given by

$$A_\gamma(k, \theta, \theta') := \frac{|k|}{2\pi i} \Theta[\gamma \cdot (\theta' - \widehat{k})] A(|k|, \theta, \theta')$$

and $k = |k|\widehat{k}$.

We wish to establish a relation between the various Fredholm determinants, and we begin by using (6.6),

$$
\begin{aligned}
1 - \widehat{F}_\gamma^+ V &= 1 - G_0^+ V - B_\gamma^0 V = (1 - G_0^+ V)[1 - (1 - G_0^+ V)^{-1} B_\gamma^0 V] \\
&= (1 - G_0^+ V)(1 - B_\gamma^+ V),
\end{aligned}
$$

where B_γ^+ is the operator whose kernel is given by

$$B_\gamma^+(k, x, y) = \frac{i}{4\pi^2} \int_{\mathbb{R}^3} dp\, \delta(k^2 - p^2) \Theta[\gamma \cdot (p - k)] \psi^+(|p|, \widehat{p}, x) e^{-ip \cdot y}.$$

For modified Fredholm determinants there is readily seen to be the following relation for products,

$$\det_2[(1 - A)(1 - B)] = \det_2(1 - A) \det_2(1 - B) e^{\operatorname{tr}[(1-A)B]}.$$

Therefore,

$$D_\gamma = \Delta \det(1 - B_\gamma^+ V) e^{\operatorname{tr}(V B_\gamma^0)}, \tag{7.2}$$

where D_γ is the modified Fredholm determinant of the Faddeev integral equation (6.16) and Δ is that of the Lippmann-Schwinger equation (1.3). The trace in the exponential is easily computed to be

$$\beta_\gamma(\mu, s) := \operatorname{tr}(V B_\gamma^0) = \frac{i}{4\pi}(k - s)\langle V \rangle,$$

where $k = \sqrt{\mu^2 + s^2}$ and $\langle V \rangle := \int_{\mathbb{R}^3} dx\, V(x)$. Furthermore,

$$\mathrm{tr}(V B_\gamma^+)^n = \mathrm{tr}(\mathsf{A}_\gamma)^n.$$

Consequently,

$$\det(\mathbb{1} - B_\gamma^+ V) = \det(\mathbb{1} - \mathsf{A}_\gamma) = \mathsf{d}_\gamma.$$

We have therefore the following relation between the various Fredholm determinants:

$$D_\gamma(\mu, s) = \varDelta(k)\mathsf{d}_\gamma(\mu, s)e^{\frac{i}{4\pi}(k-s)\langle V\rangle}. \tag{7.3}$$

This equation implies that, since for potentials in the class $\mathcal{W}$ (see Definition 1.4.1) the Fredholm determinant $\varDelta$ cannot vanish on the real axis (except at the origin), the values of $k \in \mathbb{R}^3$ for which the Fredholm determinant d_γ of (6.51) vanishes are exactly those values, and only those values, for which the function D_γ vanishes.

Lemma 7.1.1. *The exceptional points, if any, of the integral equation (6.51) coincide with the real exceptional points of the integral equation (6.16) (except possibly at the origin).*

The Fredholm determinant $\varDelta$ may be obtained from the S matrix by means of equation (1.74) and its analyticity in $\mathbb{C}^+$. As was shown below the generalized Levinson Theorem in Section 1.5, the phase 2δ of $\det S(k)$, $k \in \mathbb{R}$, can be chosen so that $\lim_{k\to\infty} \log \det S(k) = -ik\langle V\rangle/2\pi$. Therefore by Lemmas 1.4.2 and 1.74,

$$\eta(k) := \arg \varDelta(k) = -\delta(k) - \frac{k}{4\pi}\langle V\rangle \tag{7.4}$$

if we choose $\eta(+\infty) = 0$. Since $\varDelta$ is the boundary value of an analytic function that is holomorphic in $\mathbb{C}^+$ with zeros on the imaginary axis at $k = i\kappa_n$ if $E_n = -\kappa_n^2$ is an eigenvalue and since it satisfies Lemma 1.4.2, it can be explicitly expressed in terms of its phase by

$$\varDelta(k) = \lim_{\epsilon\downarrow0} \prod_n \left(1 + \frac{\kappa_n^2}{k^2}\right) \exp\left[\frac{1}{\pi} \int_0^\infty dk'^2 \frac{\arg \varDelta(k')}{k'^2 - k^2 - i\epsilon}\right]. \tag{7.5}$$

The product contains each eigenvalue as many times as its degeneracy. Thus the real number $\langle V\rangle$ can be obtained from the asymptotic form of $\det S$, and the Fredholm determinant $\varDelta$ can be computed from the S matrix. Furthermore, the Fredholm determinant d_γ of (6.51) can also be calculated from the scattering amplitude. It therefore follows from (7.2) that *for all $\mu \in \mathbb{R}_+$, $s \in \mathbb{R}$ the Fredholm determinant $D_\gamma(\mu, s)$ can be computed from a knowledge of the bound-state eigenvalues and of the scattering amplitude.*

Once the analytic function $D_\gamma(\mu, s)$ is known for real s, the number of its zeros in $\mathbb{C}^+$ can be determined. Let us define

$$\sigma := \arg D_\gamma, \qquad \zeta := \arg \mathsf{d}_\gamma$$

in such a way that $\sigma(\mu, +\infty) = 0$. Then

$$\sigma(\mu, s) = \zeta(\mu, s) + \eta(k) + \frac{k - s}{4\pi}\langle V\rangle,$$

$$\zeta(\mu, -s) = -\zeta(\mu, s) - 2\eta(k) - \frac{k}{2\pi}\langle V\rangle \quad (\mathrm{mod}\, 2\pi)$$

and $\zeta(\mu, +\infty) = 0$. If we further define $\sigma(\mu, -s) = -\sigma(\mu, s)$ then σ has a discontinuity $2\pi n(\mu)$ at $s = 0$, where n is an integer, and $2\pi n(\mu) = \sigma(\mu, 0-) - \sigma(\mu, 0+) = -2\sigma(\mu, 0+) = -2\zeta(\mu, 0+) - 2\eta(\mu) - \frac{\mu}{2\pi}\langle V\rangle$. It follows from the analyticity of $D_y(\mu, s)$ as a function of s, using the argument principle, that n equals the number of zeros of D_y in $\mathbb{C}^+$ and thus equals the number of exceptional points of (6.16) in $\mathbb{C}^+$ if there are no real exceptional points for that particular value of μ. (See the proof of Theorem 6.3.3.) Finally we use (7.4) to arrive at the following analogue of Levinson's theorem.

Lemma 7.1.2. *If $V \in \mathscr{W}$ and μ is such that the integral equation (6.16) has no real exceptional points, then the number $n(\mu)$ of exceptional points in $\mathbb{C}^+$ is related to the phases ζ of the Fredholm determinant d_y of (6.51) and 2δ of $\det S$ by*

$$\pi n(\mu) = \delta(\mu) - \zeta(0+, \mu).$$

Here δ and ζ have been chosen so that $\lim_{k\to\infty}[\delta(k) + \frac{k}{4\pi}\langle V\rangle] = 0$ and $\zeta(\mu, \infty) = 0$.

The positions and multiplicities of the zeros of $D_y(\mu, s)$ in $\mathbb{C}^+$ can, in principle, be determined from its values on the real line, for example by means of the Fourier transform of $1/D_y - 1$ with respect to the variable s.

7.2 A Gel'fand-Levitan Procedure

In this section we shall discuss the problem of finding the potential in the Schrödinger equation on the assumption that the starting point is the given function h_y and bound-state information. This may be regarded as an interesting problem in its own right, even though the function h_y does not describe scattering and has no known physical significance. If the aim is to solve the inverse scattering problem then, of course, h_y has to be found first from the scattering amplitude, as described in the previous section.

The underlying idea of the Gel'fand-Levitan procedure was discussed in Chapter 3, and it needs only certain technical modifications in the present context.

The first step is to define a solution Φ_y of the Schrödinger equation that has no poles in $\mathbb{C}^+$. For $k, x \in \mathbb{R}^3$

$$\widehat{\Phi}_y(k, x) = \Phi_y(k_\perp, s, x) := D_y(\mu, s)\varphi_y(k_\perp, s, x), \tag{7.6}$$

where D_y is the modified Fredholm determinant of (6.16). The function Φ_y is holomorphic in $\mathbb{C}^+$ and, by Lemma 6.2.4, when multiplied by $e^{-ik\cdot x}$, it approaches unity as $|s| \to \infty$ in $\mathbb{C}^+$. Consequently it has a Povzner-Levitan representation of the form

$$\Phi_y(k_\perp, s, x) = e^{ik\cdot x} - \int_{y\cdot x}^{\infty} dt\, K_y(x, k_\perp, t)e^{ist}, \tag{7.7}$$

or, more conveniently,

$$\widehat{\Phi}_y(k, x) = e^{ik\cdot x} - \int_{\mathbb{R}^3} dy\, K_y(x, y)\Theta[\gamma \cdot (y - x)]e^{ik\cdot y}, \tag{7.8}$$

where

$$K_\gamma(x, y) = \frac{1}{(2\pi)^2} \int_{\mathbb{R}^2} dk_\perp \, \mathbf{K}_\gamma(x, k_\perp, y_0) e^{-ik_\perp \cdot y_\perp}.$$

The inverse of (7.8) is given by

$$K_\gamma(x, y) = \frac{1}{(2\pi)^3} \int_{\mathbb{R}^3} dk \, [e^{ik \cdot x} - \widehat{\Phi}_\gamma(k, x)] e^{-ik \cdot y}. \tag{7.9}$$

The symmetries (6.27) and (6.22) lead to the symmetry

$$\overline{\Phi_\gamma(k_\perp, s, x)} = \Phi_\gamma(-k_\perp, -s, x), \tag{7.10}$$

or

$$\overline{\widehat{\Phi}_\gamma(k, x)} = \widehat{\Phi}_\gamma(-k, x). \tag{7.11}$$

Equation (7.9) then shows that the kernel K_γ is real:

$$\overline{K_\gamma(x, y)} = K_\gamma(x, y). \tag{7.12}$$

The kernels $\mathbf{K}_\gamma$ and K_γ may be distributions, but their most important property is that their support lies in the region $\gamma \cdot (y - x) \geq 0$, i.e., that they are *upper triangular* with respect to the direction γ. We also define the kernel

$$U_\gamma(x, y) := \delta(x - y) - K_\gamma(x, y) \Theta[\gamma \cdot (y - x)], \tag{7.13}$$

so that

$$\widehat{\Phi}_\gamma(k, x) = \int_{\mathbb{R}^3} dy \, U_\gamma(x, y) e^{ik \cdot y}. \tag{7.14}$$

If (7.8) is inserted in the Schrödinger equation then one finds, after integrating by parts, that K_γ must, for $\gamma \cdot (y - x) > 0$, satisfy the hyperbolic partial differential equation

$$(\Delta_x - \Delta_y) K_\gamma = V(x) K_\gamma, \tag{7.15}$$

and on the hyperplane $\gamma \cdot (y - x) = 0$ it must satisfy the boundary condition

$$2 \frac{\partial}{\partial x_0} K_\gamma(x_\perp, x_0, y_\perp, x_0) = V(x) \delta(x_\perp - y_\perp), \tag{7.16}$$

where $x_0 = \gamma \cdot x$, $x_\perp = x - \gamma x_0$, $y_\perp = y - \gamma x_0$, which may also be written

$$2\delta[\gamma \cdot (y - x)] \gamma \cdot (\nabla_x + \nabla_y) K_\gamma(x, y) = V(x) \delta(x - y).$$

In terms of $\mathbf{K}_\gamma$ these equations read, for $t > \gamma \cdot x = x_0$

$$\left(\Delta - \frac{\partial^2}{\partial t^2} + k_\perp^2 \right) \mathbf{K}_\gamma = V(x) \mathbf{K}_\gamma, \tag{7.17}$$

$$2 \frac{\partial}{\partial x_0} \mathbf{K}_\gamma(x_\perp, x_0, k_\perp, x_0) = V(x) e^{ik_\perp \cdot x_\perp}. \tag{7.18}$$

Lemma 7.2.1. *The system (7.15), (7.16) has at most one solution.*

Proof. We prove this result by proving that the system (7.17), (7.18) has no more than one solution. The lemma then follows by Fourier transformation. The proof for (7.17), (7.18), on the other hand, is identical to that of Lemma 2.4.1. □

7.2.1 Completeness

The relation (6.47) between the Faddeev solution φ_γ^+ and the scattering solution ψ^+ allows us to write down a completeness relation for the former. This, however, requires that we also deal with the bound states.

If one inserts (7.6), (6.47), and (6.55) in the completeness relation (1.45) one obtains a completeness relation for the functions Φ_γ,

$$\frac{1}{(2\pi)^3} \int_0^\infty d|k|\,|k|^2 \int_{S^2 \times S^2} d\theta d\theta'\, \Phi_\gamma(|k|\theta, x) M_\gamma(|k|, \theta, \theta') \overline{\Phi_\gamma(|k|\theta', y)}$$

$$+ \sum_{\kappa_n} \sum_{a,b=1}^N u_{\kappa_n}^a(x) d_{ab}^{\kappa_n} u_{\kappa_n}^b(y) = \delta(y - x), \tag{7.19}$$

where

$$M_\gamma(|k|, \theta, \theta') \;:=\; \frac{\left[\Delta_{-\gamma}^\dagger(|k|)\Delta_{-\gamma}(|k|)\right](\theta, \theta')}{\widehat{D}_\gamma(|k|\theta)\overline{\widehat{D}_\gamma(|k|\theta')}}$$

$$:=\; \delta(\theta, \theta') + m_\gamma(|k|, \theta, \theta'). \tag{7.20}$$

When $\mu \to 0$, the function φ_γ approaches the scattering solution ψ, $\varphi_\gamma(0, s, x) = \psi(s, \gamma, x)$. What is more, since by the definition of the differential operator $D_\gamma(\theta) = \frac{1}{|\theta||\gamma|}(\theta\gamma\nabla_\theta)$, where (abc) is the triple vector product and ∇_θ is the gradient with respect to θ,

$$D_\gamma(k_\perp)e^{ik\cdot x} = \frac{i}{\mu}(k\gamma x)e^{ik\cdot x}$$

and

$$D_{k_\perp}(\gamma)e^{ik\cdot x} = -\frac{is}{\mu}(k\gamma x)e^{ik\cdot x},$$

it follows from a comparison of the solutions of (1.3) and (6.16) and the use of (6.24) that

$$\lim_{\mu\to 0}[D_\gamma(k_\perp)]^m \Phi_\gamma(k_\perp, s, x) = \Delta(s)[D_{k_\perp}(\gamma)]^m \psi(s, \gamma, x)(-s)^{-m},$$

where Δ is the modified Fredholm determinant of (1.3). Now let $s \to i\kappa_n$, where $-\kappa_n^2$ is an L^2-eigenvalue of the Schrödinger equation. If this eigenvalue is N-fold degenerate then $\Delta(s)$ has an N-fold zero at $s = i\kappa_n$ and ψ has a simple pole there. Thus if $N > 1$ it is necessary to take $N - 1$ derivatives with respect to s to arrive at a non-zero result. With the definition

$$\Delta_n^N := \left.\frac{\partial^N \Delta(s)}{\partial s^N}\right|_{s=i\kappa_n}$$

then, according to (1.31), the right-hand side goes over into

$$\frac{\Delta_n^N}{N}(i\kappa_n)^{-m} \sum_{a,b=1}^N Y_{\kappa_n k_\perp}^{a(m)}(-\gamma) d_{ab}^{\kappa_n} u_{\kappa_n}^b(x),$$

in which expression the functions $Y_{\kappa_n \xi}^{a(m)}$ were defined in (1.39) and the $u_{\kappa_n}^b(x)$ are bound-state eigenfunctions that satisfy (1.27). Thus,

$$\lim_{\mu \to 0} [D_\gamma(k_\perp)]^m \Phi_\gamma^{(N-1)}(k_\perp, i\kappa_n, x)$$

$$= \frac{\Delta_n^N}{N} \sum_{a,b=1}^N Y_{\kappa_n k_\perp}^{a(m)}(-\gamma) d_{ab}^{\kappa_n} u_{\kappa_n}^b(x)(i\kappa_n)^{-m}, \qquad (7.21)$$

where $\Phi_\gamma^{(n)}$ denotes the nth derivative of Φ_γ with respect to s.

Owing to the triangular nature of the kernel K_γ the representations (7.7) and (7.8) may be analytically continued as functions of s into $\mathbb{C}^+$:

$$\Phi_\gamma(k_\perp, s, x) = e^{ik_\perp \cdot x_\perp + is\gamma \cdot x} - \int_{\mathbb{R}^3} dy\, K_\gamma(x, y) \Theta[\gamma \cdot (y - x)] e^{ik \cdot y},$$

which will be written

$$\Phi_\gamma(k_\perp, s, x) = \int_{\mathbb{R}^3} dy\, U_\gamma(x, y) e^{ik_\perp \cdot y_\perp + is\gamma \cdot y}$$

for $s \in \mathbb{C}^1$ as well. As $\mu \to 0$ and a bound-state point $s = i\kappa_n$ is approached, we use (7.21) to obtain

$$\frac{\Delta_n^N}{N} \sum_{a,b=1}^N Y_{\kappa_n k_\perp}^{a(m)}(-\gamma) d_{ab}^{\kappa_n} u_{\kappa_n}^b(x)(i\kappa_n)^{-m}$$

$$= \int_{\mathbb{R}^3} dy\, U_\gamma(x, y) [D_\gamma(k_\perp)]^m e^{ik_\perp \cdot y_\perp} \Big|_{k_\perp=0} (i\gamma \cdot y)^{N-1} e^{-\kappa_n \gamma \cdot y}$$

$$= i^m \int_{\mathbb{R}^3} dy\, U_\gamma(x, y) [(\widehat{k_\perp \gamma y_\perp})]^m (i\gamma \cdot y)^{N-1} e^{-\kappa_n \gamma \cdot y}.$$

Let us, therefore, set up N linear algebraic equations for the N real functions $\alpha_{\gamma \kappa_n}^a(x)$, $a = 1, \ldots, N$:

$$\frac{\Delta_n^N}{N} \sum_{a,b=1}^N Y_{\kappa_n k_\perp}^{a(m)}(-\gamma) d_{ab}^{\kappa_n} \alpha_{\gamma \kappa_n}^a(x)$$

$$= (-\kappa_n)^m [(\widehat{k_\perp \gamma x_\perp})]^m (i\gamma \cdot x)^{N-1} e^{-\kappa_n \gamma \cdot x}, \qquad (7.22)$$

with $m = 0, \ldots, N - 1$. The degeneracies, the character functions, and the matrix d_{κ_n} are assumed to be known. If the starting point is the scattering amplitude and the eigenvalue $-\kappa^2$ is *normal* (see Definition 1.2.5) then they can all be determined by Lemma 2.3.8. Otherwise they must be given or determined from h_γ. By Definition 1.2.5 and Corollary (1.2.6), the system (7.22) can be solved uniquely if the eigenvalue κ_n is normal, and thus the functions $\alpha_{\gamma \kappa_n}^a(x)$ can be determined. As is clear from (7.22) they are exponentials $e^{-\kappa_n x_0}$, $x_0 = \gamma \cdot x$, multiplied by x_0^{N-1} and by polynomials in $x_\perp = x - x_0 \gamma$ of degree N, if N is the degeneracy of the eigenvalue $-\kappa_n^2$.

Lemma 7.2.2. *If the potential satisfies the hypotheses of Lemma 1.2.4 and it is normal by Definition 1.2.5, then the functions $\alpha_{\gamma \kappa_n}^a(x)$ can be determined from the given scattering amplitude by solving equations (7.22).*

As a result of (7.21), (7.22), and of the analytic continuation of (7.14) we have the following representation of the bound-state eigenfunctions

$$u^a_{\kappa_n}(x) = \int_{\mathbb{R}^3} dy\, U_\gamma(x,y)\alpha^a_{\gamma\kappa_n}(y). \tag{7.23}$$

Even though the functions $\alpha^a_{\gamma\kappa_n}(x)$ increase exponentially, the upper triangularity of U_γ assures that the integral in (7.23) converges.

The completeness relation (7.19), together with (7.23), (7.14), and the reality of U_γ may now be written in the form

$$\int_{\mathbb{R}^3\times\mathbb{R}^3} dx'\, dy'\, U_\gamma(x,x')W_\gamma(x',y')U_\gamma(y,y') = \delta(x-y), \tag{7.24}$$

where the kernel W_γ is given by $W_\gamma = W^c_\gamma + W^d_\gamma$,

$$W^c_\gamma(x,y) := \frac{1}{(2\pi)^3}\int_0^\infty d|k|\,|k|^2 \int_{S^2\times S^2} d\theta d\theta'\, M_\gamma(|k|,\theta,\theta')e^{i|k|(\theta\cdot x - \theta'\cdot y)}, \tag{7.25}$$

$$W^d_\gamma(x,y) := \sum_{\kappa_n}\sum_{a,b=1}^N \alpha^a_{\gamma\kappa_n}(x)d^{\kappa_n}_{ab}\alpha^b_{\gamma\kappa_n}(y). \tag{7.26}$$

Lemma 7.2.3. *The operators whose kernels are $W^c_\gamma(x,y)$ and $W_\gamma(x,y)$ are positive definite.*

Proof. As in the proof of Lemma 4.1.1, $\int dy\, W^c_\gamma(x,y)f(y) = 0$ implies

$$\int_{S^2} d\theta'\, \Delta_{-\gamma}(|k|,\theta,\theta')f(|k|\theta') = 0$$

for almost all $|k| \in \mathbb{R}_+$. But then (6.54) implies that $f = 0$. Since W^c_γ is obviously semi-definite, this proves that it is definite. Because W^d_γ is semi-definite, it follows that W_γ is positive-definite. $\square$

7.2.2 The Povzner-Levitan Kernel

It is clear from the procedure given in Chapter 3 that it is necessary to construct the inverse of the Povzner-Levitan kernel U_γ. The argument here is completely analogous to that of Section 3.2. The upper triangularity of U_γ with respect to the direction γ implies that if the kernel K_γ is sufficiently well behaved the inverse of U_γ exists and is also upper triangular with respect to γ:

$$U^{-1}_\gamma(x,y) = \delta(x-y) + K^i_\gamma(x,y)\Theta[\gamma\cdot(y-x)], \tag{7.27}$$

where K^i_γ satisfies the quasi-Volterra equations

$$K^i_\gamma(x,y) = K_\gamma(x,y)\Theta[\gamma\cdot(y-x)] + \int_{\mathbb{R}^3} dz\, K^i_\gamma(x,z)K_\gamma(z,y)\Theta[\gamma\cdot(y-z)],$$

$$K^i_\gamma(x,y) = K_\gamma(x,y)\Theta[\gamma\cdot(y-x)] + \int_{\mathbb{R}^3} dz\, K_\gamma(x,z)K^i_\gamma(z,y)\Theta[\gamma\cdot(z-x)].$$

Inversion of (7.14) and use of (7.27) leads to the equation

$$\widehat{\Phi}_\gamma(k, x) = e^{ik\cdot x} - \int_{\mathbb{R}^3} dy\, K_\gamma^i(x, y)\Theta\left[\gamma\cdot(y-x)\right]\widehat{\Phi}_\gamma(k, y). \tag{7.28}$$

Insertion of this equation in the Schrödinger equation then shows that K_γ^i must satisfy the hyperbolic equation

$$(\Delta_x - \Delta_y)K_\gamma^i = -V(y)K_\gamma^i \tag{7.29}$$

in the region $\gamma\cdot(y-x) > 0$, and on the hyperplane $\gamma\cdot(y-x) = 0$ it must satisfy the boundary condition

$$2\frac{\partial}{\partial x_0}K_\gamma^i(x_\perp, x_0; y_\perp, x_0) = -V(x)\delta(x_\perp - y_\perp). \tag{7.30}$$

In view of equation (7.23), the kernel $U_\gamma(x, y)$ defines an operator U_γ whose domain is the space $\mathscr{H}_\gamma$ which consists of $L^2(\mathbb{R}^3)$ augmented by the functions $\alpha_{\gamma\kappa_n}^a(x)$ and whose range is all of $L^2(\mathbb{R}^3)$. Its inverse U_γ^{-1} will therefore be regarded as well defined on all of L^2, and its range is $\mathscr{H}_\gamma$. Equations (7.14) and (7.23) are equivalent to the operator statement that

$$HU_\gamma = U_\gamma H_0, \tag{7.31}$$

where $H_0 = -\Delta$, $H = -\Delta + V$.

7.2.3 The Gel'fand-Levitan-Faddeev Equation

Write equation (7.24) symbolically

$$U_\gamma W_\gamma \widetilde{U}_\gamma = \delta$$

and the Fourier theorem in the form

$$W^0 = \delta,$$

where

$$W^0(x, y) = \frac{1}{(2\pi)^3}\int_0^\infty d|k|\,|k|^2\int_{\mathbb{S}^2} d\theta\, e^{i|k|\theta\cdot(x-y)},$$

or

$$U_\gamma W_\gamma = \widetilde{U}_\gamma^{-1}, \qquad U_\gamma W^0 = U_\gamma, \tag{7.32}$$

and subtract. The result is

$$K_\gamma + \widetilde{K}_\gamma' = (\mathbb{1} - K_\gamma)g_\gamma, \tag{7.33}$$

where the kernel g_γ is given by

$$
\begin{aligned}
g_\gamma(x, y) : \;&= \; W_\gamma(x, y) - W^0(x, y)\\
&= \; \frac{1}{(2\pi)^3}\int_0^\infty d|k|\,|k|^2\int_{\mathbb{S}^2\times\mathbb{S}^2} d\theta\, d\theta'\, m_\gamma(|k|, \theta, \theta')e^{i|k|(\theta\cdot x - \theta'\cdot y)}\\
&\quad + \sum_{\kappa_n}\sum_{a,b}\alpha_{\gamma\kappa_n}^a(x)d_{ab}^{\kappa_n}\alpha_{\gamma\kappa_n}^b(y),
\end{aligned}
\tag{7.34}
$$

m_γ being defined by (7.20) in terms of the spectral function M_γ. Since both K_γ and K'_γ are upper triangular, (7.33) reads for $\gamma \cdot (y - x) > 0$

$$K_\gamma(x, y) = g_\gamma(x, y) - \int_{\mathbb{R}^3} dz\, \Theta\,[\gamma \cdot (z - x)] K_\gamma(x, z) g_\gamma(z, y), \tag{7.35}$$

which is Faddeev's generalization of the Gel'fand-Levitan equation. On the other hand, subtracting $W^0 = \delta$ from $W_\gamma = U_\gamma^{-1}\tilde{U}_\gamma^{-1}$ gives a generalized nonlinear Gel'fand-Levitan equation, for $\gamma \cdot (y - x) > 0$

$$K_\gamma^i(x, y) = g_\gamma(x, y) - \int_{\mathbb{R}^3} dz\, \Theta\,[\gamma \cdot (z - y)] K_\gamma^i(x, z) K_\gamma^i(y, z). \tag{7.36}$$

Little is known about the solvability of these equations. However, as in Chapter 3, it follows from the positive definiteness of the kernel W_γ, Lemma 7.2.3, that the spectrum of g_γ lies to the right of -1; thus the linear Gel'fand-Levitan-Faddeev equation has at most one solution. Note that in this integral equation the range of integration is not compact, in contrast to that in equation (4.8).

Remark 1. It is remarkable that the kernel and inhomogeneity of (7.35) refer only to the continuous spectrum and the bound states of the Schrödinger equation; they contain no specific information about the exceptional set $\mathscr{E}$ other than what is indirectly contained in the function D_γ.

Remark 2. Note that the functions $\alpha_{\gamma\kappa}^a(x)$ are independent of $x_\perp = x - \gamma\gamma \cdot x$. As a result the part of $g_\gamma(x, y)$ that comes from the bound states, when taken by itself, is very badly behaved both as an integral kernel and as a function of y on $\mathbb{R}^3$ in the inhomogeneity of (7.35). (The variable x is a parameter in that equation.) The only way in which (7.35) can, nevertheless, have some convenient mathematical properties when there are bound states, is for the continuum-part and the bound-state part to combine in a fortuitous manner.

If one inserts the expansion (7.34) of g_γ in (7.35) and uses (7.8) and (7.23), one obtains an expansion for K_γ:

$$K_\gamma(x, y)$$
$$= \frac{1}{(2\pi)^3} \int_0^\infty d|k|\, |k|^2 \int_{S^2 \times S^2} d\theta\, d\theta'\, m_\gamma(|k|, \theta, \theta') \widehat{\Phi}_\gamma(|k|\theta, x) e^{-i|k|\theta' \cdot y}$$
$$+ \sum_{\kappa_n} \sum_{a,b} u_{\kappa_n}^a(x) d_{ab}^{\kappa_n} \alpha_{\gamma\kappa_n}^b(y). \tag{7.37}$$

Together with (7.16) this gives a representation of the potential.

Suppose, then, that a spectral function M_γ is given and that the Gel'fand-Levitan-Faddeev equation (7.35) has been solved. Does that lead to a solution of the Schrödinger equation, and do the corresponding solutions obtained by (7.8) and (7.23) form a complete set with the spectral function M_γ?

The answer to the first question is found by applying the operator $(\Delta_x - \Delta_y)$ to (7.35), using the equation $(\Delta_x - \Delta_y)g_\gamma(x, y) = 0$, and integrating twice by parts. The result is that the kernel $\Lambda_\gamma(x, y) := (\Delta_x - \Delta_y)K_\gamma(x, y)$ solves the integral equation

$$\Lambda_\gamma(x, y) = \int_{\mathbb{R}^2} dz_\perp \, V_\gamma(x_\perp, z_\perp, x_0) g_\gamma(z_\perp, x_0, y)$$

$$- \int_{\mathbb{R}^3} dz \, \Theta\left[\gamma \cdot (z - x)\right] \Lambda_\gamma(x, z) g_\gamma(z, y),$$

where

$$V_\gamma(x_\perp, z_\perp, x_0) := 2\frac{\partial}{\partial x_0} K_\gamma(x_\perp, x_0; z_\perp, x_0). \tag{7.38}$$

Therefore it follows from the uniqueness of the solution of (7.35) that

$$\Lambda_\gamma(x, y) = \int_{\mathbb{R}^2} dz_\perp \, V_\gamma(x_\perp, z_\perp, x_0) K_\gamma(z_\perp, x_0; y).$$

In other words, for $\gamma \cdot (y - x) > 0$, K_γ satisfies the partial differential equation

$$(\Delta_x - \Delta_y)K_\gamma(x, y) = \int_{\mathbb{R}^2} dz_\perp \, V_\gamma(x_\perp, z_\perp, x_0) K_\gamma(z_\perp, x_0; y), \tag{7.39}$$

where V_γ is given by (7.38). If this equation is used in (7.8) and (7.23), one finds that $\widehat{\Phi}_\gamma$ satisfies the Schrödinger equation with the nonlocal potential V_γ in the same sense in which that potential enters in (7.39), and so do the functions $u^a_{\kappa_n}(x)$. Thus the solution of the Gel'fand-Levitan-Faddeev equation will always lead, via (7.8) and (7.23), to a Schrödinger equation, but the corresponding potential will generally be nonlocal. Put in another way, if U_γ is defined by (7.13), where K_γ solves (7.35), then

$$H_\gamma = U_\gamma H_0 U_\gamma^{-1}$$

will be of the form $H_\gamma = H_0 + V_\gamma$, where V_γ is the operator whose kernel is given by (7.38). If the solution of (7.35) is such that it is "miraculously" of the form of equation (7.16) then the potential is local.

In order to show that the solution of (7.35) leads to the correct spectral function we proceed as in the proof of Lemma 4.1.2. Let $K_\gamma(x, y) = 0$ for $\gamma \cdot (y - x) < 0$ and let it solve (7.35) for $\gamma \cdot (y - x) > 0$; define $\widetilde{K}''_\gamma := g_\gamma - K_\gamma - K_\gamma g_\gamma$, so that K''_γ is upper triangular. Then one finds that $K_\gamma - K''_\gamma + K_\gamma K''_\gamma = K_\gamma g_\gamma + g_\gamma \widetilde{K}_\gamma - g_\gamma + K_\gamma + \widetilde{K}_\gamma - K_\gamma g_\gamma \widetilde{K}_\gamma - K_\gamma \widetilde{K}_\gamma$. The left-hand side is upper triangular and the right-hand side is symmetric in x and y since g_γ is symmetric. Hence both must vanish, which implies that $(\mathbb{1} - K_\gamma)(\mathbb{1} + K''_\gamma) = \mathbb{1}$, i.e., $K''_\gamma = K^i_\gamma$. Thus we have (7.33), which is equivalent to (7.24), and (7.36) follows.

Theorem 7.2.4. *Suppose that g_γ is given in the form (7.34) and K_γ satisfies the Gel'fand-Levitan-Faddeev equation (7.35). Then the functions Φ_γ and $u^a_{\kappa_n}$ defined by (7.8) and (7.23) satisfy the Schrödinger equation with the potential (7.38), which will generally be nonlocal, except when it reduces to the form (7.16). Furthermore, the functions Φ_γ and $u^a_{\kappa_n}$ form a complete set with the spectral data in (7.34).*

It should be noted that in this procedure the direction γ is fixed, and the resulting potential (7.38) will generally depend on this direction; V_γ is local with respect to the direction γ. From the point of view of the inverse spectral problem, the circle is closed: spectral data $\longrightarrow$ potential $\longrightarrow$ spectral data, and no "miracle" is required.

On the other hand, from the point of view of the inverse scattering problem the circle is *not* necessarily closed. There is no known reason to believe that if one started with a given scattering amplitude, found h_γ by solving (6.16), constructed M_γ by (7.20), solved (7.35), constructed Φ_γ by (7.8) and ψ by (6.47), that the resulting scattering amplitude would necessarily coincide with that given initially.

Suppose now that (7.35) is solved for all directions γ and that the potential (7.38) comes out to be independent of γ. Since (7.38) is local in the direction γ this implies that V is, in fact, *local*. Thus Φ_γ and $u^a_{\gamma\kappa_n}$ satisfy the Schrödinger equation with a local potential. The question is whether there is a way of assuring from the start that the potential constructed is local. The answer was given by Faddeev.

Lemma 7.2.5. (Faddeev) *If the scattering amplitude A is such that the kernel $D_\gamma h_\gamma$ [in which h_γ is the solution of (6.51) and D_γ is constructed by means of (7.2)] is in $\mathcal{T}$, then the potential constructed from (7.35) (if that equation has a solution) is local.*

This lemma will be proved in Subsection 7.2.4.

With regard to the inverse scattering problem we are now in the following position. Suppose that one starts with a given scattering amplitude A; then h_γ is found by solving (6.51) and the Fredholm determinant D_γ is found by (7.2); thus we have Δ_γ by (6.48) and (6.49), and hence we know M_γ and g_γ by (7.20) and (7.34). We now solve (7.35) and obtain K_γ and V_γ; this gives us Φ_γ by (7.8), which solves the Schrödinger equation and satisfies the completeness relation with the spectral weight M_γ. (See Theorem 7.2.4.) Assume that a "miracle" happens and the constructed $V = V_\gamma$ is *local*.

Now use the potential V in the Schrödinger equation and solve the direct problem, thus calculating the scattering amplitude A' and the Faddeev solution φ'_γ. Find $D'_\gamma, \Phi'_\gamma, \Delta'_\gamma, M'_\gamma$ [from (7.20) and Δ'_γ] and g'_γ from (7.34). Solving (7.35) with g'_γ as input, we obtain K'_γ. On what grounds can one assert that all the primed quantities equal the unprimed ones?

Lemma 7.2.1 assures us that, since both K_γ and K'_γ solve the same hyperbolic system (7.15),(7.16), $K_\gamma = K'_\gamma$. It follows from (7.8) that therefore $\Phi_\gamma = \Phi'_\gamma$, and since M_γ and M'_γ are their spectral weights, $M_\gamma = M'_\gamma$.

The function D_γ, is uniquely related to the kernel K_γ by the requirement that for $x \in \mathscr{C}^+_{\gamma,k}$ [see (6.15) and Definition 6.1.1]

$$\lim_{|x|\to\infty} \int_{\mathbb{R}^3} dy\, K_\gamma(x,y) e^{ik\cdot(y-x)} = D_\gamma(k) - 1$$

because of (6.34) and (7.6). Therefore $D_\gamma = D'_\gamma$. If there is assurance that the factorization in (7.20) is unique, then $\Delta_\gamma = \Delta'_\gamma$ and hence $h_\gamma = h'_\gamma$. Finally, one regards (6.51) as a quasi-Volterra equation for A [or one uses (6.52) to construct A] and thus one finds that $A' = A$, and the circle is closed.

The missing step is supplied by the following lemma, which will be proved in Subsection 7.2.4.

Lemma 7.2.6. *The factorization (7.20) of $\mathcal{M}_\gamma := D_\gamma M_\gamma \overline{D}_\gamma^{-1}$ into two functions Δ_γ and $\Delta_\gamma^\dagger$ with the upper triangularity property embodied in (6.48) and (6.49) and the symmetries (6.37) and (6.40), is unique.*

We summarize our results in the following.

Theorem 7.2.7. *Suppose that $A \in \mathcal{A}$ (see Definition 1.5.15) is given and that the solution $\widehat{h}_\gamma$ of the integral equation (6.51) is such that $D_\gamma \widehat{h}_\gamma \in \mathcal{T}$ (see Definition 6.4.1), where D_γ is constructed from A. Suppose further that the Gel'fand-Levitan-Faddeev equation (7.35) is solvable and leads, by (7.38), to a local potential V. Then the solution K_γ of (7.35) leads, via (7.8), to a regular solution Φ_γ and, via (6.47), to a scattering solution ψ^+ both of which satisfy the Schrödinger equation with the potential V. Furthermore the scattering amplitude obtained from ψ^+ by (1.2) is equal to the initially given function A.*

The potential thus obtained can, in fact, be explicitly computed from the function h_γ. The analyticity and asymptotic properties of the function H_γ defined below Definition 6.4.1 are embodied in the Hilbert-transform relation

$$f(s) = \frac{1}{i\pi} \mathscr{P} \int_{-\infty}^{\infty} ds' \frac{f(s')}{s' - s}$$

for the function

$$f(s) := H_\gamma(k'_\perp, k_\perp, s) + \frac{1}{4\pi} \int_{\mathbb{R}^3} dx\, V(x) e^{i(k_\perp - k'_\perp) \cdot x}.$$

Here $\mathscr{P}$ denotes Cauchy's principal value. Thus we have the "dispersion relation"

$$
\begin{aligned}
f(s) + f(-s) &= \frac{1}{i\pi} \mathscr{P} \int_{-\infty}^{\infty} ds' \frac{f(s') - f(-s')}{s' - s} \\
&= \frac{2}{i\pi} \mathscr{P} \int_{0}^{\infty} ds'\, s' \frac{f(s') - f(-s')}{s'^2 - s^2},
\end{aligned}
$$

or

$$
\begin{aligned}
H_\gamma(k'_\perp, k_\perp, s) \;+\;& H_\gamma(k'_\perp, k_\perp, -s) \\
+\; \frac{2i}{\pi} \mathscr{P} \int_0^\infty ds'\, s' & \frac{H_\gamma(k'_\perp, k_\perp, s') - H_\gamma(k'_\perp, k_\perp, -s')}{s'^2 - s^2} \\
=\;& -\frac{1}{2\pi} \int_{\mathbb{R}^3} dx\, V(x) e^{i(k_\perp - k'_\perp) \cdot x}.
\end{aligned} \tag{7.40}
$$

If, for a given value of $\mu = |k_\perp| = |k'_\perp|$, there are real exceptional points then H_γ has poles on the real axis and the principal value in the above integral applies to them as well.

In view of Faddeev's Lemma and Lemma 6.4.2 this equation may be regarded both as a necessary and sufficient condition for a local potential to exist, and as an explicit formula for that potential in terms of the exceptional points and the function h_γ (which are combined in H_γ). If the left-hand side of (7.40) is independent of γ and s and is a function of $k_\perp - k'_\perp$ only, then a local potential exists and its Fourier transform is given by it in the sense of the right-hand side.

Conclusion. The inverse scattering problem may thus be solved directly by solving the relatively simple integral equation (6.51) and then computing the Fourier transform of the potential by (7.40). Equation (6.51) is an integral equation at fixed $|k|$ and γ, but it has to be solved for all $|k| \in \mathbb{R}_+$ and all $\gamma \in S^2$. For an arbitrarily given function $A \in \mathscr{A}$ the solution of (6.51) will generally not produce, via (7.40), a function that is independent of γ and s and depends on $k_\perp - k'_\perp$ only. That admissible amplitudes A must do so is the analogue of the miracle in Chapter 2. One might attempt to short-circuit part of this miracle by using the function h_γ as the initial input and compute V by (7.40) and the scattering amplitude A by means of (6.51). In that case, however, it will require another miracle for the thus calculated function A to be independent of γ.

7.2.4 Proofs

Proof of Lemma 7.2.5. The proof proceeds by means of differentiation with respect to γ, using the operator $D_\xi(\gamma)$ defined in (1.38), which for now we shall simply denote by D. Let us apply it to equation (6.51). If that equation has a unique solution (i.e., we are not at an exceptional point) then it follows from the result of such differentiation that

$$D\widehat{h}_\gamma(|k|, \theta', \theta) = \int_{S^2} d\theta'' \, \widehat{h}_\gamma(|k|, \theta', \theta'')\widehat{h}_\gamma(|k|, \theta'', \theta)\omega_\gamma^\xi(|k|, \theta'', \theta),$$

where

$$\omega_\gamma^\xi(|k|, \theta'', \theta) := \frac{|k|}{2\pi i}\left((\theta'' - \theta)\gamma\xi\right)\delta\left[\gamma \cdot (\theta'' - \theta)\right]$$

and (abc) is the triple vector product. It therefore follows from (6.48) and (6.49) that

$$D\Delta_{-\gamma}(|k|, \theta', \theta) = \int_{S^2} d\theta'' \, \Delta_{-\gamma}(|k|, \theta', \theta'')\omega_\gamma^\xi(|k|, \theta'', \theta)\widehat{h}_{-\gamma}(|k|, \theta'', \theta).$$

Writing

$$\lambda_\gamma^\xi(|k|, \theta', \theta) \; := \; \omega_\gamma^\xi(|k|, \theta', \theta)\widehat{h}_\gamma(|k|, \theta', \theta)$$

$$= \; \frac{|k|}{2\pi i}\left((\theta' - \theta)\gamma\xi\right)\delta\left[\gamma \cdot (\theta' - \theta)\right]h_\gamma(k'_\perp, s, k_\perp, s),$$

this may be expressed in operator notation on S^2 as

$$D\Delta_{-\gamma} = -\Delta_{-\gamma}\lambda_{-\gamma}^\xi.$$

Therefore by (6.40) and (6.37), with $\mathscr{M}_\gamma := \widetilde{\Delta}_{-\gamma}\Delta_{-\gamma}$,

$$D\mathscr{M}_\gamma = \lambda_\gamma^\xi \mathscr{M}_\gamma - \mathscr{M}_\gamma \lambda_{-\gamma}^\xi. \tag{7.41}$$

The next step is to calculate the derivative of the Fredholm determinant:

$$D \log D_\gamma(k) \; = \; D \log \det_2[\mathbb{1} - \widehat{F}_\gamma(k)V]$$

$$= \; -\mathrm{tr}[\widehat{F}_\gamma V (\mathbb{1} - \widehat{F}_\gamma V)^{-1}D\widehat{F}_\gamma V].$$

From (6.6)

$$\begin{aligned}
D\widehat{F}_\gamma(k,x) &= DB_\gamma^0(k,x) \\
&= \frac{i}{(2\pi)^2} \int_{\mathbb{R}^3} dp\, e^{ip\cdot x}\delta(p^2-k^2)D\Theta[\gamma\cdot(p-k)] \\
&= \frac{i}{(2\pi)^2} \int_{\mathbb{R}^3} dp\, e^{ip\cdot x}\delta(p^2-k^2)\delta[\gamma\cdot(p-k)](\gamma\xi(p-k)).
\end{aligned}$$

The use of (6.13) and of the integral equation (6.16) therefore leads to

$$\begin{aligned}
D\log D_\gamma(k) &= -\frac{i}{(2\pi)^2} \int_{\mathbb{R}^3} dp\, (\gamma\xi(p-k))\delta(p^2-k^2)\delta[\gamma\cdot(p-k)] \\
&\quad \times \int_{\mathbb{R}^3\times\mathbb{R}^3} dx\, dy\, \widehat{F}(k,x-y)V(y)\widehat{\varphi}_\gamma(p,y)e^{-ip\cdot x}V(x) \\
&= \frac{i}{(2\pi)^2} \int_{\mathbb{R}^3} dp\, (\gamma\xi(p-k))\delta(p^2-k^2)\delta[\gamma\cdot(p-k)] \\
&\quad \times \int_{\mathbb{R}^3} dx\, e^{-ip\cdot x}V(x)[e^{ip\cdot x}-\widehat{\varphi}_\gamma(p,x)] \\
&= \frac{i}{\pi} \int_{\mathbb{R}^3} dp\, (\gamma\xi(p-k))\delta(p^2-k^2)\delta[\gamma\cdot(p-k)] \\
&\quad \times \left[\widehat{h}_\gamma(|k|,\hat{p},\hat{p}) + \frac{\langle V\rangle}{4\pi}\right] \\
&= -\int_{S^2} d\theta\, \omega_\gamma^\xi(|k|,\theta,\hat{k}) \left[\widehat{h}_\gamma(|k|,\theta,\theta) + \frac{\langle V\rangle}{4\pi}\right].
\end{aligned}$$

Now from (7.20), (6.22), and (6.23),

$$DM_\gamma = \frac{1}{D_\gamma}D\mathcal{M}_\gamma\frac{1}{D_{-\gamma}} - (D\log D_\gamma)M_\gamma - M_\gamma(D\log D_{-\gamma}).$$

Thus we obtain from (7.25):

$$DW_\gamma^c = \Pi_\gamma^{\xi-}W_\gamma^c - W_\gamma^c\Pi_{-\gamma}^{\xi+},$$

where the operators now act on functions on $\mathbb{R}^3$, and the kernels of $\Pi_\gamma^{\xi\pm}$ are

$$\begin{aligned}
\Pi_\gamma^{\xi\pm}(x,y) &= \frac{1}{(2\pi)^3} \int_0^\infty d|k|\,|k|^2 \int_{S^2\times S^2} d\theta\, d\theta'\, [\lambda_\gamma^\xi(|k|,\theta,\theta') \\
&\quad \pm\delta(\theta,\theta')D\log D_\gamma(|k|\theta)]e^{i|k|(\theta\cdot x-\theta'\cdot y)} \\
&= \frac{1}{(2\pi)^3} \int_0^\infty d|k|\,|k|^2 \int_{S^2\times S^2} d\theta d\theta'\, \omega_\gamma^\xi(|k|,\theta,\theta')e^{-i|k|\theta'\cdot y} \\
&\quad \times \left[\widehat{h}_\gamma(|k|,\theta,\theta')e^{i|k|\theta\cdot x} \mp \left[\widehat{h}_\gamma(|k|,\theta',\theta') + \frac{\langle V\rangle}{4\pi}\right]e^{i|k|\theta'\cdot x}\right] \\
&= \frac{i}{(2\pi)^3\pi} \int_{\mathbb{R}^2\times\mathbb{R}^2} dk_\perp\, dk'_\perp\, \delta(k_\perp^2-k'^2_\perp)((k'_\perp-k_\perp)\gamma\xi) \\
&\quad \times e^{-ik'_\perp\cdot y_\perp} \int_{-\infty}^\infty ds\, e^{is(x_0-y_0)}\left[h_\gamma(k_\perp,s,k'_\perp,s)e^{ik_\perp\cdot x_\perp}\right]
\end{aligned}$$

$$\mp \left[h_\gamma(k_\perp, s, k_\perp, s) + \frac{\langle V \rangle}{4\pi} \right] e^{ik'_\perp \cdot x_\perp}.$$

Thus, if $h_\gamma(k_\perp, s, k'_\perp, s)$ has the properties stated in Lemma 6.4.2 then the kernels $\Pi_\gamma^{\xi\pm}$ are upper triangular with respect to the direction γ.

We next set up an equation that is analogous to the first equation in (7.32) but using only W_γ^c,

$$U_\gamma^c W_\gamma^c = \tilde{U}_\gamma^{c-1},$$

and require that $U_\gamma^c - \mathbb{1}$ be lower triangular. This defines U_γ^c by means of a Gel'fand-Levitan-Faddeev equation that has the same structure as (7.35), and whose solution is unique because W_γ^c is positive definite (Lemma 7.2.3).

Now apply D to the equation $U_\gamma W_\gamma^c \tilde{U}_\gamma = \delta - U_\gamma W_\gamma^d \tilde{U}_\gamma$, using the fact that the right-hand side [in which W_γ^d is the part of W_γ defined by (7.26)] is independent of γ:

$$(DU_\gamma) W_\gamma^c \tilde{U}_\gamma + U_\gamma (DW_\gamma^c) \tilde{U}_\gamma + U_\gamma W_\gamma^c (D\tilde{U}_\gamma) = 0,$$

which becomes

$$(U_\gamma^{-1} DU_\gamma + \Pi_\gamma^{\xi-}) W_\gamma^c = W_\gamma^c (\Pi_{-\gamma}^{\xi+} - D\tilde{U}_\gamma \tilde{U}_\gamma^{-1});$$

this implies

$$\Gamma_\gamma W_\gamma^c = \tilde{U}_\gamma^{c-1} (\Pi_{-\gamma}^{\xi+} - D\tilde{U}_\gamma \tilde{U}_\gamma^{-1}),$$

where $\Gamma_\gamma := U_\gamma^c U_\gamma^{-1} (DU_\gamma + U_\gamma \Pi_\gamma^{\xi-})$. The right-hand side of this equation is lower triangular with respect to γ, and Γ_γ is upper triangular. Therefore the equation implies that Γ_γ satisfies the homogeneous form of the Gel'fand-Levitan-Faddeev equation (7.35). Uniqueness of the solution of (7.35) thus implies that $\Gamma_\gamma = 0$, or, since upper triangularity prevents U_γ^c and U_γ^{-1} from having nontrivial null-spaces,

$$DU_\gamma = -U_\gamma \Pi_\gamma^{\xi-}.$$

This result is now used in equation (7.31):

$$\begin{aligned} DH_\gamma &= (DU_\gamma) H_0 U_\gamma^{-1} - U_\gamma H_0 U_\gamma^{-1} (DU_\gamma) U_\gamma^{-1} \\ &= U_\gamma (H_0 \Pi_\gamma^{\xi-} - \Pi_\gamma^{\xi-} H_0) U_\gamma^{-1}. \end{aligned}$$

The δ-distibution in $\Pi_\gamma^{\xi-}$ makes it "on the energy shell" and hence it commutes with H_0. Consequently,

$$DH_\gamma = 0$$

and thus the potential is local.   □

Proof of Lemma 7.2.6. Suppose that there are two such factorizations, $\Delta_\gamma^\dagger \Delta_\gamma = \Delta_\gamma'^\dagger \Delta_\gamma'$, or by (6.55), $\Delta'_{-\gamma} \tilde{\Delta}_\gamma = \Delta'_\gamma \tilde{\Delta}_{-\gamma}$, which, according to (6.48), reads

$$q'_{-\gamma} - \tilde{\tilde{q}}_\gamma - \frac{|k|}{2\pi i} q'_{-\gamma} \tilde{\tilde{q}}_\gamma = q'_\gamma - \tilde{\tilde{q}}_{-\gamma} - \frac{|k|}{2\pi i} q'_\gamma \tilde{\tilde{q}}_{-\gamma}.$$

By (6.49), (6.37), and (6.40) this equation says that for $\gamma \cdot (\theta - \theta') > 0$,

$$\widehat{h}'_\gamma(|k|, \theta, \theta')$$
$$= \widehat{h}_\gamma(|k|, \theta, \theta') + \frac{|k|}{2\pi i} \int_{\gamma \cdot \theta' < \gamma \cdot \theta'' < \gamma \cdot \theta} d\theta'' \, \widehat{h}'_\gamma(|k|, \theta, \theta'') \widehat{h}_\gamma(|k|, \theta'', \theta'),$$

which is a quasi-Volterra equation for $\widehat{h}'_\gamma$. Hence its solution is unique. According to (6.58), $\widehat{h}'_\gamma = \widehat{h}_\gamma$ solves it. Hence that is the only solution and $\Delta'_\gamma = \Delta_\gamma$ is the only factorization.   □

7.3 A Marchenko Procedure

In order to utilize a Marchenko method for the solution of the inverse scattering problem we return to Faddeev's integral equation (6.16) for solutions of the Schrödinger equation with $s \in \mathbb{C}^+$ and with $s \in \mathbb{C}^-$. These are related to one another by (6.25) or (6.26). Using equations (6.55) and (6.47) and the fact that ψ^+ does not depend on γ, one then obtains the relation

$$\widehat{\varphi}^-_\gamma(k, x) = \int_{S^2} d\hat{k}' \, \mathcal{M}_\gamma(|k|, \hat{k}, \hat{k}') \widehat{\varphi}^+_\gamma(k', x), \tag{7.42}$$

in which $\hat{k} = k/|k|$, $\hat{k}' = k'/|k|$, $|k'| = |k|$ and $\mathcal{M}_\gamma(|k|, \hat{k}', \hat{k})$ is the kernel of the operator

$$\mathcal{M}_\gamma = \widetilde{\Delta}_{-\gamma} \overline{\Delta}_{-\gamma}, \tag{7.43}$$

Δ_γ being defined by (6.48). This operator is positive definite.

Let us define the functions

$$\widehat{\chi}^\pm_\gamma(k, x) := \widehat{\varphi}^\pm_\gamma(k, x) e^{-ik \cdot x}$$

and the kernel

$$\widehat{\Omega}_{\gamma x}(k, p) = \Omega_{\gamma x}(k_\perp, s, p) := \mathcal{M}_\gamma(|k|, \hat{k}, \hat{p}) e^{ix \cdot (p-k)},$$

where $|k| = |p|$, $\hat{k} = k/|k|$, $\hat{p} = p/|p|$, $s = \gamma \cdot k$, $k_\perp = k - s\gamma$. Then equation (7.42) becomes

$$\widehat{\chi}^-_\gamma(k, x) = \int_{\mathbb{R}^3} dp \, \frac{2}{|k|} \delta(k^2 - p^2) \widehat{\Omega}_{\gamma x}(k, p) \widehat{\chi}^+_\gamma(p, x). \tag{7.44}$$

The functions χ_γ share the analytic properties of the solutions φ_γ of (6.16) and according to Lemma 6.2.4 they approach unity as $|s| \to \infty$. Therefore equation (7.44) gives rise to what may be called a *nonlocal* Riemann-Hilbert problem because, in contrast to the usual situation, equation (7.44) contains an integral over the variable as a function of which χ^+ and χ^- have analytic continuations into $\mathbb{C}^+$ and $\mathbb{C}^-$, respectively. We shall discuss only the case without exceptional points of (6.16). The generalization to that with exceptional points in $\mathbb{C}^+$ would follow the lines of Section 2.3.2. If there are real exceptional points then the s-integrations have to be understood in the sense of Cauchy's principal value.

If (6.16) has no exceptional points then the function $\chi^+_\gamma(k_\perp, s, x) := \widehat{\chi}^+_\gamma(k, x)$ (in our usual notation convention) is holomorphic in $\mathbb{C}^+$ and thus the function

$$\lambda_\gamma(k_\perp, t, x) := \frac{1}{2\pi} \int_{-\infty}^{\infty} ds\, e^{-ist} [\chi^+_{-\gamma}(k_\perp, s, x) - 1] \tag{7.45}$$

vanishes for $t < 0$. We take the Fourier transform of equation (7.44), defining

$$R_{\gamma x}(k_\perp, p_\perp, t, t')$$

$$= \frac{1}{\pi} \int_{-\infty}^{\infty} ds \int_{-\infty}^{\infty} ds'\, \delta(k_\perp^2 - p_\perp^2 + s^2 - s'^2) \frac{1}{|k|} e^{i(st+s't')} [\hat{\Omega}_{\gamma x}(k, p) - \delta(\hat{k}, \hat{p})], \tag{7.46}$$

where $k = k_\perp + \gamma s$, $p = p_\perp + \gamma s'$, and we obtain

$$\lambda_\gamma(k_\perp, t, x) = \lambda_{-\gamma}(k_\perp, -t, x) + \int_{\mathbb{R}^2} dp_\perp\, R_{\gamma x}(k_\perp, p_\perp, t, 0)$$

$$+ \int_{\mathbb{R}^2} dp_\perp \int_{-\infty}^{\infty} dt'\, R_{\gamma x}(k_\perp, p_\perp, t') \lambda_{-\gamma}(p_\perp, t', x).$$

It then follows that for $t > 0$

$$\lambda_\gamma(k_\perp, t, x) = \int_{\mathbb{R}^2} dp_\perp\, R_{\gamma x}(k_\perp, p_\perp, t, 0)$$

$$+ \int_{\mathbb{R}^2} dp_\perp \int_0^{\infty} dt'\, R_{\gamma x}(k_\perp, p_\perp, t, t') \lambda_{-\gamma}(p_\perp, t', x). \tag{7.47}$$

This is another generalization of the Marchenko equation. The nonlocal character of the Riemann-Hilbert problem manifests itself in the fact that the kernel $R_{\gamma x}$ is not a function of $t + t'$ only but depends on t and t' separately. One iteration of (7.47) makes it into an equation for λ_γ alone. Nothing further is known about the properties of this integral equation. Specifically, neither existence nor uniqueness of a solution have been established.

Insertion of the Fourier transform (7.45) in the Schrödinger equation leads, in the same manner as before, to a hyperbolic equation for λ_γ together with the boundary condition

$$V(x) = 2\gamma \cdot \nabla \lambda_\gamma(k_\perp, 0+, x). \tag{7.48}$$

This equation may be used to calculate the potential. That the right-hand side, in spite of appearances, must come out to be independent of both γ and $k_\perp$, is the analogue of the miracle in Chapter 2. The discussion of the relation between the Riemann-Hilbert problem and the generalized Marchenko equation in Chapter 2 is applicable here and it will not be repeated.

Remark. If there are exceptional points in $\mathbb{C}^+$, they will have to be handled analogously to the bound states in Chapter 2. It is not sufficient to multiply φ_γ by the function D_γ and to proceed with the function Φ_γ as in the last section. The solution of the Riemann-Hilbert problem will in general not be unique unless the *operator* solution is such that neither it nor its inverse has poles in $\mathbb{C}^+$, and removing poles by a scalar factor will not achieve that.

7.4 The $\bar{\partial}$-approach

In the two methods based on Faddeev's integral equation discussed so far the solution was considered a function of the complex variable s and as such to live entirely in the upper half of the complex s-plane. We will now regard this solution as a function of the vector $k \in \mathbb{C}^3$. In that domain a distinction between upper and lower half-planes cannot be made. Thus in the $\bar{\partial}$-approach the functions $\hat{\varphi}$ and $\hat{\xi}$ are taken to live in all of $\mathbb{C}^3$. Let us return to Faddeev's integral equation (6.16) with $k \in \mathbb{C}^3 \backslash \mathbb{R}^3$. In that case the Green's functions $\hat{F}$ and $\widehat{\mathscr{F}}$ are well-defined function of k, and unless k is exceptional, so are the solutions $\hat{\varphi}(k, x)$ and $\hat{\xi}(k, x)$.

The set of exceptional points $k \in \mathbb{C}^3$ will now be denoted by $\mathscr{E}$. Since because of (6.12) the modified Fredholm determinant of (6.16) has the symmetry $\hat{D}(k + p) = \hat{D}(k)$ for $p \in \mathbb{R}^3$, $p^2 + 2p \cdot k = 0$ (which is simply the statement that $D_y(|k_\perp|, s)$ is invariant under rotations of $k_\perp$ about y) the exceptional set for a fixed value of $\gamma = \Im k/|\Im k|$ consists of a (possibly infinite) number of two-dimensional manifolds. Therefore $\mathscr{E}$ consists of a (possibly infinite) number of four-dimensional manifolds in $\mathbb{C}^3$. The symmetries (6.2) and (6.9) imply that $\hat{D}$ has the symmetry $\hat{D}(-k) = \hat{D}(k)$. It follows from this and (6.22) that whenever $k \in \mathscr{E}$ then $-k$, $\bar{k}$, and $-\bar{k}$ are also in $\mathscr{E}$. R. G. Novikov and G. M. Henkin have proved further measure-theoretic details of $\mathscr{E}$ [HN88].

The functions $\hat{\varphi}$ and $\hat{\xi}$ are analytic in k for all $k \in \mathbb{C}^3 \backslash (\mathbb{R}^3 \cup \mathscr{E})$. As $k \to \mathbb{R}^3$ in the sense that $\Im k = \gamma \epsilon$, $\gamma \in \mathbb{R}^3$, $|\gamma| = 1$, we have $\lim_{\epsilon \downarrow 0} \widehat{\mathscr{F}}(k + i\gamma\epsilon, x) := \widehat{\mathscr{F}}_\gamma^+(k, x) = \mathscr{F}_\gamma^+(k_\perp, s, x)$, where $s = \gamma \cdot k$, $k_\perp = k - \gamma s$.

The $\bar{\partial}$ of the Green's function $\widehat{\mathscr{F}}$ is easily calculated from (5.20) and (6.11),

$$\bar{\partial}_j \widehat{\mathscr{F}}(k, x) := \frac{\partial}{\partial \bar{k}_j} \widehat{\mathscr{F}}(k, x) = -\frac{1}{(2\pi)^2} \int_{\mathbb{R}^3} dp\, e^{ip \cdot x} p_j \delta_{\mathbb{C}}(p^2 + 2p \cdot k).$$

Therefore, by differentiating (6.16) and using (6.35), one obtains

$$\int_{\mathbb{R}^3} dy\, [\mathbb{1} - \widehat{\mathscr{F}}(k)V](x, y)\bar{\partial}_j \hat{\xi}(k, y)$$

$$= \int_{\mathbb{R}^3} dy\, [\bar{\partial}_j \widehat{\mathscr{F}}(k, x - y)]V(y)\hat{\xi}(k, y)$$

$$= -\frac{1}{(2\pi)^2} \int_{\mathbb{R}^3} dp\, e^{ip \cdot x} \delta_{\mathbb{C}}(p^2 + p \cdot k) \int_{\mathbb{R}^3} dy\, V(y)e^{-ip \cdot y}\hat{\xi}(k, y).$$

As a consequence of (6.12) we have for $p \in \mathbb{R}^3$, $p^2 + 2p \cdot k = 0$,

$$\int_{\mathbb{R}^3} dy\, [\mathbb{1} - \widehat{\mathscr{F}}(k)V]^{-1}(x, y)e^{ip \cdot y} = \hat{\xi}(k + p, x)e^{ip \cdot x}, \tag{7.49}$$

and we obtain

$$\bar{\partial}_j \hat{\xi}(k, x) = -\frac{1}{(2\pi)^2} \int_{\mathbb{R}^3} dp\, p_j \delta_{\mathbb{C}}(p^2 + 2p \cdot k)\hat{\xi}(k + p, x)e^{ip \cdot x}$$

$$\times \int_{\mathbb{R}^3} dy\, V(y)e^{-ip \cdot y}\hat{\xi}(k, y) + \sigma,$$

where σ is the sum of the pole contributions. If Conjecture 6.3.2 holds, then

$$\sigma = \pi \sum_n \text{Res}_n(x)\delta_{\mathbb{C}}(k_j - \kappa_{nj}),$$

where the residues are solutions of the homogeneous version of (6.16). The y-integral can be replaced by

$$-\frac{1}{4\pi} \int_{\mathbb{R}^3} dy\, V(y)e^{-ip\cdot y}\widehat{\xi}(k, y) := h(p, k).$$

The function $h(p, k)$ is well defined for $p \in \mathbb{R}^3$ and $k \in \mathbb{C}^3\backslash(\mathbb{R}^3 \cup \mathscr{E})$; it is, in fact, an analytic function of k there. As $\mathfrak{I}k = \gamma\epsilon$, $|\gamma| = 1$, $\epsilon > 0$, approaches zero, $h(p, k) \to h_\gamma(p, k)$, which is such that for $|k| = |p|$, $h_\gamma(p-k, k) = h_\gamma(p_\perp, p\cdot\gamma, k_\perp, k\cdot\gamma) = \widehat{h}_\gamma(|k|, \hat{p}, \hat{k})$, where h_γ and $\widehat{h}_\gamma$ are the functions defined by (6.35) and (6.36) but expressed in different variables. Thus,

$$\overline{\partial}_j\widehat{\xi}(k, x) = \frac{1}{\pi} \int_{\mathbb{R}^3} dp\, p_j\delta_{\mathbb{C}}(p^2 + 2p\cdot k)\widehat{\xi}(k + p, x)e^{ip\cdot x}h(p, k) + \sigma. \tag{7.50}$$

In order to avoid the pole contributions it is convenient to define the function

$$\widehat{\Xi}(k, x) := \widehat{D}(k)\widehat{\xi}(k, x), \tag{7.51}$$

where $\widehat{D}(k)$ is the modified Fredholm determinant of the Faddeev integral equation (6.16). This function has all the analyticity and asymptotic properties of $\widehat{\xi}$ as a function of k, except that it has no singularities on $\mathscr{E}$. The first thing we need, then, is $\overline{\partial}_j\widehat{D}(k) := \partial\widehat{D}/\partial\overline{k}_j$:

$$\overline{\partial}_j \log \widehat{D}(k) = \overline{\partial}_j \log \det_2[\mathbb{1} - \widehat{\mathscr{F}}(k)V] = -\text{tr}[\widehat{\mathscr{F}}V(\mathbb{1} - \widehat{\mathscr{F}}V)^{-1}\overline{\partial}_j\widehat{\mathscr{F}}V].$$

A short computation, using (6.12), shows that

$$\begin{aligned}\overline{\partial}_j \log \widehat{D}(k) &= \frac{1}{(2\pi)^2} \int_{\mathbb{R}^3} dp\, p_j\delta_{\mathbb{C}}(p^2 + 2p\cdot k) \\ &\quad \times \int_{\mathbb{R}^3\times\mathbb{R}^3} dx\, dy\, V(x)V(y)\widehat{\mathscr{F}}(k + p, x - y)\widehat{\xi}(k + p, y).\end{aligned}$$

Equation (6.19) and the definition of h then leads to

$$\begin{aligned}\overline{\partial}_j \log \widehat{D}(k) \\ &= \frac{1}{(2\pi)^2} \int_{\mathbb{R}^3} dp\, p_j\delta_{\mathbb{C}}(p^2 + 2p\cdot k) \int_{\mathbb{R}^3} dx\, V(x)[\widehat{\xi}(k + p, x) - 1] \\ &= -\frac{1}{\pi} \int_{\mathbb{R}^3} dp\, p_j\delta_{\mathbb{C}}(p^2 + 2p\cdot k) \left[h(0, k + p) + \frac{\langle V \rangle}{4\pi}\right], \tag{7.52}\end{aligned}$$

in which $\langle V \rangle := \int_{\mathbb{R}^3} dx\, V(x)$. This expression is then used in

$$\overline{\partial}_j\widehat{\Xi}(k, x) = \widehat{D}(k)\overline{\partial}_j\widehat{\xi}(k, x) + \widehat{\Xi}(k, x)\overline{\partial}_j \log \widehat{D}(k).$$

Equation (6.12) implies that for $p^2 + 2p\cdot k = 0$, $\widehat{D}(k+p) = \widehat{D}(k)$. Thus one obtains

$$\bar{\partial}_j \hat{\Xi}(k, x) = \frac{1}{\pi} \int_{\mathbb{R}^3} dp \, p_j \delta_{\mathbb{C}}(p^2 + 2p \cdot k)$$

$$\times \left[\hat{\Xi}(k + p, x) e^{ip \cdot x} h(p, k) - \left[h(0, k + p) + \frac{\langle V \rangle}{4\pi} \right] \hat{\Xi}(k, x) \right]. \quad (7.53)$$

The pole terms σ have been eliminated by the function $\hat{D}(k)$.

The next step is the use of equation (5.19) to arrive at a reconstruction of $\hat{\Xi}$ from its $\bar{\partial}$:

$$\hat{\Xi}(k, x) = 1 + \frac{1}{\pi^2} \int_{-\infty}^{\infty} \int_{-\infty}^{\infty} \frac{dk'_{Rj} \, dk'_{Ij}}{k_j - k'_j} \int_{\mathbb{R}^3} dp \, p_j \delta_{\mathbb{C}}(p^2 + 2p \cdot k')$$

$$\times \left[\hat{\Xi}(k' + p, x) e^{ip \cdot x} h(p, k') - \left[h(0, k' + p) + \frac{\langle V \rangle}{4\pi} \right] \hat{\Xi}(k', x) \right]. \quad (7.54)$$

This is to be regarded as an integral equation for the solution $\hat{\Xi}$ of the reduced Schrödinger equation (6.18).

The singularities on $\mathscr{E}$ may also be eliminated from the function $h(p, k)$ by multiplying it by $\hat{D}(k)$. Thus we define

$$H(p, k) := \hat{D}(k) h(p, k) = -\frac{1}{4\pi} \int_{\mathbb{R}^3} dx \, V(x) e^{-ip \cdot x} \hat{\Xi}(k, x). \quad (7.55)$$

It is now an easy task to calculate the $\bar{\partial}$ of H by means of (7.50). The result is

$$\bar{\partial}_j H(p, k) = \frac{1}{\pi \hat{D}(k)} \int_{\mathbb{R}^3} dp' \, p'_j \delta_{\mathbb{C}}(p'^2 + 2p' \cdot k) \left[H(p - p', k + p') H(p', k) \right.$$

$$\left. - H(p, k) \left[H(0, k + p') + \hat{D}(k) \frac{\langle V \rangle}{4\pi} \right] \right]. \quad (7.56)$$

It should be noted that the contents of Lemma 6.4.2 can be stated in terms of H simply as

$$\lim_{|k| \to \infty} H(p, k) = -\frac{1}{4\pi} \int_{\mathbb{R}^3} dx \, V(x) e^{-ip \cdot x}. \quad (7.57)$$

The following theorem has been proved by R. G. Novikov and G. M. Henkin [NH86,87b, and HN88]:

Theorem 7.4.1. *If the potential is in the Schwartz class C_∞^∞ then H, defined by (7.55), is in the Schwartz class C_∞^∞ and satisfies the nonlinear equation (7.56). Moreover, the corresponding function $\hat{\Xi}$ satisfies (7.53). Conversely, if $H \in C_\infty^\infty$ and it satisfies (7.56), then there exists a unique potential in C_∞^∞ such that (7.55) holds, and the Fourier transform $\hat{V}$ of this potential is given by (7.57).*

For the proof, see [NH86,87b]. The same authors also proved the following important theorem:

Theorem 7.4.2. *If the potential has sufficiently small norm*

$$\|V\| = \sup_x |V(x)e^{-\alpha|x|}|$$

for some $\alpha > 0$, then it is uniquely determined by a given scattering amplitude at a fixed $|k| > 0$.

[NH87b, Theorem 4.9.]

7.5 Notes

7.1 Equation (7.3) was given by [Ne74] and, with a minor error, by [Fa74]. Lemma 7.1.2 is new.

7.2 This section is based on [Fa74] and [Ne74a and 74b]. It should be noted that the effect of the presence of the functions D_γ and $\overline{D}_\gamma$ in the spectral function (7.20) is that the kernel $m_\gamma(|k|, \theta, \theta')$ contains a Dirac distribution $\delta(\theta, \theta')$. This can be avoided by the method used in [Ne74b], in which the poles of φ_γ at the exceptional points s_n in $\mathbb{C}^+$ are removed by multiplying φ_γ by the product $P_\gamma := \prod_n (s - s_n)/(s - \bar{s}_n)$ instead of by D_γ as in (7.6). In that case $\overline{P}_\gamma = 1/P_\gamma$ for real s and hence the factor $1/P_\gamma \overline{P}_\gamma$ in (7.20) produces no Dirac distribution in m_γ because $\delta(\theta, \theta')$ in M_γ has the factor 1. However, the replacement of D_γ by P_γ makes other matters more cumbersome.

The details of Subsection 7.2.1 are new, while most of the contents of Subsections 7.2.2 and 7.2.3 come from [Fa74] and [Ne74a and 74b]. Lemma 7.2.6 and Theorems 7.2.4 and 7.2.7 are new; Faddeev's Lemma 7.2.5 was given by [Fa74], with a proof for the case without exceptional points. The proof given in Subsection 7.2.5 mostly follows and fleshes out that of Faddeev, but the part that deals with the Fredholm determinant D_γ, and hence with exceptional points, is new. Formula (7.40) was first given by [Ne85a] and [NH87a].

7.3 This section follows [Ne85a].

7.4 The use of the $\overline{\partial}$-method in the inverse scattering problem was pioneered by Nachman and Ablowitz [NA84], Beals and Coifman [BC85, 86, and 87], and Novikov and Henkin [NH86, 87a, 87b, and HN88]. This section mostly follows these authors. Henkin and Novikov have also generalized the use of the $\overline{\partial}$-method to the Schrödinger equation with a vector potential in addition to the potential V [HN88], and Nachman and Ablowitz have generalized the method to the Schrödinger equation with a time-dependent potential [NA84]. See also [Me88] for a detailed application of the $\overline{\partial}$-method to multidimensional scattering theory.

The inverse scattering problem for the Schrödinger equation with a magnetic field, as in Remark 8 of Section 1.5, was the subject of chapter V of [NH87b]; see also [HN88].

References

[AM63]   Agranovich, Z. S., and V. A. Marchenko, The Inverse Problem of Scattering Theory, Gordon and Breach, New York (1963).

[AS71]   Alsholm, P., and G. Schmidt, *Spectral and scattering theory for Schrödinger operators*, Arch. Ratl. Mech. Anal. **40** (1971) 281–311.

[AJS77]   Amrein, Werner O., Joseph M. Jauch, and Kalyan B. Sinha, Scattering Theory in Quantum Mechanics, W.A.Benjamin, Inc., Reading (1977).

[BC85]   Beals, Richard, and Ronald R. Coifman, *Multidimensional inverse scattering and nonlinear P.D.E.*, Proc. Symp. Pure Math. **43** (1985) 45–70.

[BC86]   Beals, Richard, and Ronald R. Coifman, *The d-bar approach to inverse scattering and nonlinear evolutions*, Physica D **18** (1986) 242–249.

[BC87]   Beals, Richard, and Ronald R. Coifman, *Multidimensional scattering and inverse scattering*, Yale University preprint.

[Bi61]   Birman, M. S., *The spectrum of singular boundary problems*, Mat. Sb. **55** (1961) 125–174, [English translation: Am. Math. Soc. Translations **53** (1966) 23–80].

[Ca82]   Carroll, Robert, Transmutations, Scattering Theory and Special Functions, North Holland Publishing Co., Amsterdam (1982).

[CS77]   Chadan, K., and P. C. Sabatier, Inverse Problems in Quantum Scattering Theory, Springer Verlag, New York (1977); second edition (1989).

[Ch82]   Cheney, Margaret, *Quantum mechanical scattering and inverse scattering in two dimensions*, Ph.D.Thesis, Indiana University (1982) unpublished.

[Ch84a]   Cheney, Margaret, *Inverse scattering in dimension two*, J. Math. Phys. **25** (1984) 94–107.

[Ch84b]   Cheney, Margaret, *Two-dimensional scattering: The number of bound states from scattering data*, J. Math. Phys. **25** (1984) 1449–1455.

[Ch84c]   Cheney, Margaret, *A rigorous derivation of the "miracle" identity of three-dimensional inverse scattering*, J. Math. Phys. **25** (1984) 2988–2990.

[Ch85]   Cheney, Margaret, *Two-dimensional inverse scattering: Compactness of the generalized Marchenko operator*, J. Math. Phys. **26** (1985) 743–752.

[CR85]   Cheney, Margaret, and James H. Rose, *Three-dimensional inverse scattering: High-frequency analysis of Newton's Marchenko equation*, J. Math. Phys. **26** (1985) 436–439.

[CR88a]   Cheney, Margaret, and James H. Rose, *Generalization of the Fourier transform: Implications for inverse scattering theory*, Phys. Rev. Letters **60** (1988) 1221–1224.

[CR88b]   Cheney, Margaret, and James H. Rose, *Three-dimensional inverse scattering for the wave equation: weak scattering approximation with error estimates*, Inverse Problems **4** (1988) 435–447.

[CRD87a]   Cheney, Margaret, James H. Rose, and Brian DeFacio, *Three-dimensional inverse scattering*, in Differential Equations and Mathematical Physics, Springer Verlag (1987) 46–54.

[CRD87b]   Cheney, Margaret, James H. Rose, and Brian DeFacio, *A fundamental equation of scattering theory*, SIAM J. Math. Anal. **19** (1988) 1090–1102.

[CRD87c]   Cheney, Margaret, James H. Rose, and Brian DeFacio, *A new equation of scattering theory and its use in inverse scattering*, Duke University preprint.

[DR85]   DeFacio, Brian, and James H. Rose, *Inverse scattering theory for the non-spherically-symmetric three-dimensional plasma wave equation*, Phys. Rev. A **31** (1985) 897–902.

[Dr76]   Dreyfus, Tommy, *On the number of bound states and the determinant of the scattering matrix*, Ph.D. Thesis, University of Geneva (1976), unpublished.

[Dr78a]   Dreyfus, Tommy, *The determinant of the scattering matrix and its relation to the number of eigenvalues*, J. Math. Anal. Appl. **64** (1978) 114–134.

[Dr78b]   Dreyfus, Tommy, *The number of states bound by noncentral potentials*, Helv. Physica Acta **51** (1978) 321–329.

[ER88]   Eskin, G., and J. Ralston, *The inverse backscattering problem in three dimensions*, UCLA preprint.

[Fa56]   Faddeev, L.D., *Uniqueness of the inverse scattering problem*, Vestnik Leningrad Univ. **11** No.7 (1956) 126–130 [Math. Revs. **18** 259].

[Fa65]   Faddeev, L. D., *Increasing solutions of the Schrödinger equation*, Dokl. Akad. Nauk SSSR **165** (1965) 514-517, [English translation: Soviet Physics Doklady **10** (1966) 1033–1035].

[Fa66]   Faddeev, L. D., *Factorization of the S matrix for the multidimensional Schrödinger operator*, Dokl. Akad. Nauk SSSR **167** (1966) 69-72, [English translation: Soviet Physics Doklady **11** (1966) 209–211].

[Fa71]   Faddeev, L. D., *Three-dimensional inverse problem in the quantum theory of scattering*, Academy of Sciences of the Ukarainian SSR, Kiev (1971) preprint ITP-71-106E.

[Fa74]   Faddeev, L. D., *Inverse problem of quantum scattering theory II*, Itogi Nauki i Tekhniki, Sov. Prob. Mat. **3** (1974) 93–180 [English translation: J. Soviet Math. **5** (1976) 334–396].

[GL51]   Gel'fand, I. M., and B. M. Levitan, *On the determination of a differential equation from its spectral function*, Isvest. Akad. Nauk SSSR **15** (1951) 309 [English translation: American Mathematical Society Translations **1** (1955) 253–304].

[GK86]   Gohberg, I., and M. A. Kaashoek, Constructive Methods of Wiener-Hopf Factorization, Birkhäuser Verlag, Basel (1986).

[GMP87]   Greenberg, W., C. V. M. van der Mee, and V. Protopopescu, Boundary Value Problems in Abstract Kinetic Theory, Birkhäuser Verlag, Basel (1987).

[GM86]   Grinevich, P. G., and S. V. Manakov, *Inverse scattering problem for the two-dimensional Schrödinger operator, the $\bar{\partial}$-method and nonlinear equations*, Funkt. Anal. i Ego Prol. **20** (1986) 14-24, [English translation: Funct. Anal. and Appl. **20** (1986) 94–103].

[HN88]   Henkin, G. M., and R. G. Novikov, *A multidimensional inverse problem in quantum and acoustic scattering*, Inverse Problems **4** (1988) 103–122.

[Ho66]   Hörmander, Lars, An Introduction to Complex Analysis in Several Variables, D. van Nostrand Co., Princeton, J.N. (1966).

[Ka78]   Karlsson, Bengt, *Inverse method for off-shell continuation of the scattering amplitude in quantum mechanics*, in Applied Inverse Problems, P. C. Sabatier, editor, Springer Verlag, Berlin (1978) 226–247.

[Ka59]   Kato, T., *Growth properties of of solutions of the reduced wave equation with variable coefficients*, Commun. Pure Appl. Math **12** (1959) 403–425.

[KM55]   Kay, I., and H. E. Moses, *The determination of the scattering potential from the spectral measure function. I. Continuous spectrum*, Nuovo Cimento **2** (1955) 917–961.

[KM56]   Kay, I., and H. E. Moses, *The determination of the scattering potential from the spectral measure function. II. Point eigenvalues and proper eigenfunctions*, Nuovo Cimento **3** (1956) 66–84.

[KM61a]   Kay, I., and H. E. Moses, *The determination of the scattering potential from the spectral measure function, V. The Gel'fand-Levitan equation for the three-dimensional scattering problem*, Nuovo Cimento **22** (1961) 689–705.

[KM61b]   Kay, I., and H. E. Moses, *A simple verification of the Gel'fand-Levitan equation for the three-dimensional scattering problem*, Commun. Pure and Appl. Math. **14** (1961) 435–445.

[Kr82]   Krantz, Steven G., Function Theory of Several Complex Variables, John Wiley and Sons, New York (1982).

[Ku78]   Kuroda, S. T., An Introduction to Scattering Theory, Lecture Notes Series, No. 51, Matematisk Institut, Aarhus University (1978).

[La73]   Lavine, Richard B., *Absolute continuity of positive spectrum for Schrödinger operators with long-range potentials* J. Funct. Anal. **12** (1973) 30–54.

[LN86a]   Lavine, Richard B., and Adrian I. Nachman, *On the inverse scattering transform of the n-dimensional Schrödinger operator*, in Topics in Soliton Theory and Exactly Solvable Nonlinear Equations, M. Ablowitz, editor, World Scientific Publ., Singapore (1987).

[LN86b]   Lavine, Richard B., and Adrian I. Nachman, *The Faddeev-Lippmann-Schwinger equation in multidimensional quantum inverse scattering*, in Inverse Problems: An Interdisciplinary Study, P. C. Sabatier, editor, Academic Press (1987) 169–174.

[Lu66]   Ludwig, D., *The Radon transform in Euclidean space*, Commun. Pure Appl. Math. **19** (1966) 49–81.

[Mar55]   Marchenko, V. A., *The construction of the potential energy from the phases of the scattered waves*, Dokl. Akad. Nauk SSSR **104** (1955) 695–698 [Math. Revs. **17**, p.740].

[Me87]   Melin, Anders, *Intertwining methods in multi-dimensional scattering theory*, University of Lund preprint.

[Mo56]   Moses, H. E., *Calculation of the scattering potential from reflection coefficients*, Phys. Rev. **102** (1956) 559–567.

[Mu77]   Muskelishvili, N. I., Singular Integral Equations, Noordhoff Publishing Co., Leyden (1977).

[Na86]   Nachman, Adrian, *Multidimensional inverse scattering for the time-dependent and time-independent Schrödinger equation*, presented at the Conference on Differential Equations and Mathematical Physics, Birmingham, Alabama (1986), unpublished.

[NA84]   Nachman, Adrian I., and Mark J. Ablowitz, *A multidimensional inverse scattering method*, Studies in Appl. Math, **71** (1984) 243–250.

[Ne74a]   Newton, Roger G., *The Gel'fand-Levitan method in the inverse scattering problem in quantum mechanics*, in Scattering Theory in mathematical Physics, J.A. Lavita and J.-P.Marchand, editors, D. Reidel Publishing Co., Dordrecht (1974) 193–225.

[Ne74b]   Newton, Roger G., *The three-dimensional inverse scattering problem in quantum mechanics*, invited lectures at the 1974 summer seminar on inverse problems, American Math. Soc., U.C.L.A., August, 1974, unpublished.

[Ne74c]   Newton, Roger G., *The determinantal method for bound states and resonances of three-particle systems*, Czech. J. of Physics **B24** (1974) 1195–1204.

[Ne77a]   Newton, Roger G., *Noncentral potentials: The generalized Levinson theorem and the structure of the spectrum*, J. Math. Phys. **18** (1977) 1348–1357.

[Ne77b]   Newton, Roger G., *Nonlocal interactions: The generalized Levinson theorem and the structure of the spectrum*, J. Math. Phys. **18** (1977) 1582–1588.

[Ne79]   Newton, Roger G., *New result on the inverse scattering problem in three dimensions*, Phys. Rev. Letters **43** (1979) 541–542.

[Ne80]   Newton, Roger G., *Inverse scattering.II. Three dimensions*, J. Math. Phys. **21** (1980) 1698–1715, Errata: **22**, 631 and **23**, 693.

[Ne81]   Newton, Roger G., *Inverse scattering.III. Three dimensions, continued*, J. Math. Phys. **22** (1981) 2191–2200, Errata **23**, 693.

[Ne82a]   Newton, Roger G., *Inverse scattering.IV. Three dimensions: Generalized Marchenko construction with bound states* , J. Math. Phys. **23** (1982) 2257–2265.

[Ne82b]   Newton, Roger G., *On a generalized Hilbert problem*, J. Math. Phys. **23** (1982) 2257–2265.

[Ne82c]   Newton, Roger G., Scattering Theory of Waves and Particles, Second edition, Springer Verlag, New York (1982).

[Ne83]   Newton, Roger G., *The Marchenko and Gel'fand-Levitan methods in the inverse scattering problem in one and three dimensions*, in CONFERENCE ON INVERSE SCATTERING: THEORY AND APPLICATION, J.B.Bednar et al., editors, SIAM, Philadelphia (1983) 1–74.

[Ne84]   Newton, Roger G., *Representation of the potential in the Schrödinger equation*, Phys. Rev. Letters **53** (1984) 1863–1865.

[Ne85a]   Newton, Roger G., *A Faddeev-Marchenko method for inverse scattering in three dimensions*, Inverse Problems **1** (1985) 127–132.

[Ne85b]   Newton, Roger G., *Variational principles for inverse scattering*, Inverse Problems **1** (1985) 371–380.

162   References

[Ne85c]   Newton, Roger G., *Relation between the Schrödinger equation and the plasma wave equation*, Phys. Rev. A **31** (1985) 3305–3308.

[Ne87]   Newton, Roger, G., *Some open questions in multi-dimensional inverse problems*, in Differential Equations in Mathematical Physics, I. W. Knowles and Y. Saitō, editors, Springer Verlag, New York (1987) 352–360.

[Ne88a]   Newton, Roger G., *Stability of the generalised Marchenko method*, Inverse Problems **4** (1988) 541–548.

[Ne88b]   Newton, Roger G., *A variational principle for inverse scattering*, in Theory and Applications of Inverse Problems, H. Haario, editor, Longman Scientific and Technical, Harlow (1988) 58–67.

[Ne89a]   Newton, Roger G., *Eigenvalues of the S matrix*, Phys. Rev. Lett. **62** (1989) 1811–1812.

[Ne89b]   Newton, Roger G., *The spectrum of the Schrödinger S matrix: Low energies and a new Levinson theorem*, Annals of Physics (1989), to be published.

[No86a]   Novikov, R. G., *Construction of two-dimensional Schrödinger operator with given scattering amplitude at fixed energy*, Teor. i Mat. Fiz **66** (1986) 234–240, [English translation: Theoret. Math. Phys. **66** (1986) 154–158].

[No86b]   Novikov, R. G., *Reconstruction of a two-dimensional Schrödinger operator from the scattering amplitude for fixed energy*, Funkt. Anal. i Ego Prol. **20** (1986) 90–91, [English translation: Funct. Anal. and Appl. **20** (1986) 246–248].

[NH86]   Novikov, R. G., and G. M. Henkin, *$\bar{\partial}$-equation in a multidimensional inverse scattering problem*, Institute of Physics, Krasnojarsk, preprint (in Russian).

[NH87a]   Novikov, R. G., and G. M. Henkin, *Solution of a multidimensional inverse scattering problem on the basis of generalized dispersion relations*, Dokl. Akad. Nauk SSSR **292** 814–818, [English translation: Soviet Math. Doklady **35** (1987) 153–157].

[NH87b]   Novikov, R. G., and G. M. Henkin, *The $\bar{\partial}$-equation in the multidimensional inverse scattering problem* Uspekhi Mat. Nauk **42** (1987) 93–151 [English translation: Russian Math. Surveys **42** (1987) 109–180].

[Pr69]   Prosser, R. T., *Formal solution of inverse scattering problems*, J. Math. Phys. **10** (1969) 1819–1822.

[Pr76]   Prosser, R. T., *Formal solution of inverse scattering problems II*, J. Math. Phys. **17** (1976) 1775–1779.

[Pr80]   Prosser, R. T., *Formal solution of inverse scattering problems III*, J. Math. Phys. **21** (1980) 2648–2653.

[Pr82]   Prosser, R. T., *Formal solution of inverse scattering problems IV*, J. Math. Phys. **23** (1982) 2127–2130.

[Ra87a]   Ramm, A. G., *Completeness of the products of solutions to PDE and uniqueness theorems in inverse scattering*, Inverse Problems **3** (1987) L77–82.

[Ra88a]   Ramm, A. G., *Numerical method for solving 3D inverse scattering problems*, Appl. Math. Lett., **1** (1988) 381–384.

[Ra88b]   Ramm, A. G., *Multidimensional inverse problems and completeness of the products of solutions to PDE*, J. Math. Anal. Appl. **134** (1988) 211–253.

[Ra88c]   Ramm, A. G., *Multidimensional inverse problems: Uniqueness theorems*, Appl. Math. Lett. **1** (1988) 377–380.

[RW87]   Ramm, A. G., and O. L. Weaver, *A characterisation of the scattering data in the 3D inverse scattering problem*, Inverse Problems **3** (1987) L49–52.

[RS79]   Reed, Michael, and Barry Simon, Methods of Moder Mathematical Physics III. Scattering Theory, Academic Press, New York (1979).

[RC87]   Rose, James H., and Margaret Cheney, *Self-consistent equations for variable velocity three-dimensional inverse scattering*, Phys. Rev. Letters **59** (1987) 954–957.

[RC88]   Rose, James H., and Margaret Cheney, *Generalized eigenfunction expansion for scattering in inhomogeneous three-dimensional media*, J. Math. Phys. **29** (1988) 1347–1355.

[RCD84]   Rose, James H., Margaret Cheney, and Brian DeFacio, *The connection between time- and frequency-domain three-dimensional inverse scattering methods*, J. Math. Phys. **25** (1984) 2995–3000.

[RCD85]   Rose, James, H., Margaret Cheny, and Brian Defacio, *Three-dimensional inverse scattering: Plasma and variable velocity Wave equations*, J. Math. Phys. **26** (1985) 2803–2813.
[RCD86]   Rose, James, H., Margaret Cheny, and Brian Defacio, *Determination of the wavefield from scattering data*, Phys. Rev. Letters **57** (1986) 783–786.
[Sb84]   Sabatier, Pierre C., *Well-posed questions and exploration of the space of parameters in linear and nonlinear inversion*, in Inverse Problems of Acoustic and Elastic Waves, F. Santosa *et al.*, editors, SIAM, Philadelphia (1984) 82–103.
[Sb87a]   Sabatier, Pierre C., *A few geometrical features of inverse and ill-posed problems*, in Inverse and Ill-posed Problems, H. W. Engl and C. W. Groetsch, editors, Academic Press, New York (1987).
[Sb87b]   Sabatier, Pierre C., *Remark on the three-dimensional mixed impedance potential equation*, Inverse Problems **3** (1987) L83–86.
[Sb88]   Sabatier, Pierre C., *For an impedance scattering theory*, in Nonliear Evolutions, J. Léon, editor, World Scientific Publ., Singapore (1988) 723–745.
[Sb89a]   Sabatier, Pierre C., *Three-dimensional impedance scattering theory*, in Electromagnetic and Acoustic Scattering: Detection and Inverse Problems, C. Bourrely et al., editors, World Scientific Publ., Singapore (1989).
[Sb89b]   Sabatier, Pierre C., *Green's functions for chains of Schrödinger equations*, in Proceedings of the 4th Conference on Control of Distributed Parameter Systems (Vorau, 1988), K. Kunisch et al., editors (1989).
[Sb89c]   Sabatier, Pierre C., *Perturbation theory in impedance scattering*, Montpellier preprint (1989).
[Sb89d]   Sabatier, Pierre C., *On modeling discontinuous media*, J. Math. Phys. **30** (1989), to be published.
[Sa82a]   Saitō, Yoshimi, *Some properties of the scattering amplitude and the inverse scattering problem*, Osaka J. Math. **19** (1982) 527–547.
[Sa82b]   Saitō, Yoshimi, *An inverse problem in potential theory and the inverse scattering problem*, J. Math. Kyoto University **22** (1982) 302–327.
[Sa84]   Saitō, Yoshimi, *An asymptotic behavior of the S matrix in the inverse scattering problem*, J. Math. Phys. **25** (1984) 3105–3111.
[Sa86]   Saitō, Yoshimi, *An approximation formula in the inverse scattering problem*, J. Math. Phys. **27** (1986) 1145–53.
[Sc61]   Schwinger, Julian, *On the bound states of a given potential* Proc. Natl. Acad. Sci. U.S. **47** (1961) 122–129.
[Si71]   Simon, Barry, Quantum Mechanics for Hamiltonians Defined as Quadratic Forms, Princeton University Press (1971).
[So87]   Somersalo, Erkki, *Inverse scattering for standing wave solutions of the Schrödinger equation*, J. Math. Phys. **28** (1987) 2416–2419.
[Ve67]   Vekua, N. P., Systems of Singular Integral Equations, P. Noordhoff, Groningen (1967).
[We85]   Weder, Ricardo, *Analyticity of the scattering matrix for wave propagation in crystals*, J. Math. Pures et Appl. **64** (1985), 121–148.
[We88]   Weder, Ricardo, *Multidimensional inverse scattering theory*, preprint dated August, 1988, Instituto de Investigaciones en Matematicas Applicadas y en Sistemas, Universidad Nacional Autonoma de Mexico, Mexico, D.F.
[We89]   Weder, Ricardo, *Multidimensional inverse scattering. The reconstruction problem*, preprint dated March, 1989, Instituto de Investigaciones en Matemáticas Applicadas y en Sistemas, Universidad Nacional Autónoma de México, México, D.F.
[ZK58]   Zemach, C., and A. Klein, *The Born expansion in non-relativistic quantum theory*, Nuovo Cimento **X** (1958) 1078–1087.

Index of Symbols

Subject Index